Advances in Experimental
Philosophy of Science

Advances in Experimental Philosophy

Series Editor

James Beebe, Associate Professor of Philosophy, University at Buffalo, USA

Editorial Board

Joshua Knobe, Yale University, USA
Edouard Machery, University of Pittsburgh, USA
Thomas Nadelhoffer, College of Charleston, USA
Eddy Nahmias, Neuroscience Institute at Georgia State University, USA
Jennifer Nagel, University of Toronto, Canada
Joshua Alexander, Siena College, USA

Empirical and experimental philosophy is generating tremendous excitement, producing unexpected results that are challenging traditional philosophical methods. *Advances in Experimental Philosophy* responds to this trend, bringing together some of the most exciting voices in the field to understand the approach and measure its impact in contemporary philosophy. The result is a series that captures past and present developments and anticipates future research directions.

To provide in-depth examinations, each volume links experimental philosophy to a key philosophical area. They provide historical overviews alongside case studies, reviews of current problems, and discussions of new directions. For upper-level undergraduates, postgraduates, and professionals actively pursuing research in experimental philosophy, these are essential resources.

Titles in the series include

Advances in Experimental Epistemology, edited by James R. Beebe
Advances in Experimental Moral Psychology, edited by Hagop Sarkissian and Jennifer Cole Wright
Advances in Experimental Philosophy and Philosophical Methodology, edited by Jennifer Nado
Advances in Experimental Philosophy of Aesthetics, edited by Florian Cova and Sébastien Réhault
Advances in Experimental Philosophy of Language, edited by Jussi Haukioja
Advances in Experimental Philosophy of Logic and Mathematics, edited by Andrew Aberdein and Matthew Inglis
Advances in Experimental Philosophy of Mind, edited by Justin Sytsma
Advances in Religion, Cognitive Science, and Experimental Philosophy, edited by Helen De Cruz and Ryan Nichols
Experimental Metaphysics, edited by David Rose
Methodological Advances in Experimental Philosophy, edited by Eugen Fischer and Mark Curtis

Advances in Experimental Philosophy of Science

Edited by
Daniel A. Wilkenfeld and Richard Samuels

BLOOMSBURY ACADEMIC
LONDON • NEW YORK • OXFORD • NEW DELHI • SYDNEY

BLOOMSBURY ACADEMIC
Bloomsbury Publishing Plc
50 Bedford Square, London, WC1B 3DP, UK
1385 Broadway, New York, NY 10018, USA
29 Earlsfort Terrace, Dublin 2, Ireland

BLOOMSBURY, BLOOMSBURY ACADEMIC and the Diana logo are trademarks
of Bloomsbury Publishing Plc

First published in Great Britain 2019
This paperback edition published in 2021

Copyright © Daniel A. Wilkenfeld, Richard Samuels and Contributors 2019

Daniel A. Wilkenfeld, Richard Samuels have asserted their right under the Copyright,
Designs and Patents Act, 1988, to be identified as Editors of this work.

Cover design by Catherine Wood
Cover photograph © Dieter Leistner / Gallerystock

All rights reserved. No part of this publication may be reproduced or transmitted
in any form or by any means, electronic or mechanical, including photocopying,
recording, or any information storage or retrieval system, without prior permission
in writing from the publishers.

Bloomsbury Publishing Plc does not have any control over, or responsibility for, any
third-party websites referred to or in this book. All internet addresses given in this
book were correct at the time of going to press. The author and publisher regret any
inconvenience caused if addresses have changed or sites have ceased to exist, but
can accept no responsibility for any such changes.

A catalogue record for this book is available from the British Library.

Library of Congress Cataloging-in-Publication Data
Names: Wilkenfeld, Daniel A., editor. | Samuels, Richard (Richard Ian), editor.
Title: Advances in experimental philosophy of science / edited by
Daniel A. Wilkenfeld and Richard Samuels.
Description: London; New York, NY: Bloomsbury Academic, 2019. |
Series: Advances in experimental philosophy | Includes bibliographical
references and index.
Identifiers: LCCN 2019021034 (print) | LCCN 2019022179 (ebook) | ISBN
9781350068865 (hb: alk. paper)
Subjects: LCSH: Science–Philosophy.
Classification: LCC Q175 .A2935 2019 (print) | LCC Q175 (ebook) |
DDC 501–dc23
LC record available at https://lccn.loc.gov/2019021034
LC ebook record available at https://lccn.loc.gov/2019022179

ISBN: HB: 978-1-3500-6886-5
PB: 978-1-3502-5006-2
ePDF: 978-1-3500-6887-2
eBook: 978-1-3500-6888-9

Series: Advances in Experimental Philosophy

Typeset by Deanta Global Publishing Services, Chennai, India

To find out more about our authors and books visit www.bloomsbury.com
and sign up for our newsletters.

Contents

1 Introduction *Richard Samuels and Daniel Wilkenfeld*   1

Part One Explanation and Understanding

2 Scientific Discovery and the Human Drive to Explain *Elizabeth Kon and Tania Lombrozo*   15
3 The Challenges and Benefits of Mechanistic Explanation in Folk Scientific Understanding *Frank Keil*   41

Part Two Theories and Theory Change

4 Information That Boosts Normative Global Warming Acceptance without Polarization: Toward J. S. Mill's Political Ethology of National Character *Michael Andrew Ranney, Matthew Shonman, Kyle Fricke, Lee Nevo Lamprey, and Paras Kumar*   61
5 Doubly Counterintuitive: Cognitive Obstacles to the Discovery and the Learning of Scientific Ideas and Why They Often Differ *Andrew Shtulman*   97
6 Intuitive Epistemology: Children's Theory of Evidence *Mark Fedyk, Tamar Kushnir, and Fei Xu*   122

Part Three Special Sciences

7 Applying Experimental Philosophy to Investigate Economic Concepts: Choice, Preference, and Nudge *Michiru Nagatsu*   147
8 Scientists' Concepts of Innateness: Evolution or Attraction? *Edouard Machery, Paul Griffiths, Stefan Linquist, and Karola Stotz*   172

Part Four General Considerations

9 Causal Judgment: What Can Philosophy Learn from Experiment? What Can it Contribute to Experiment? *James Woodward*   205

List of Contributors   245
Index   249

1

Introduction

Richard Samuels and Daniel Wilkenfeld

Introduction

It is hard to think of any human endeavor more transformative in its effects than science. In the space of a few hundred years, it has generated dramatic improvements in our understanding of the universe and spawned a plethora of new technologies. It pervades almost every aspect of contemporary human life: our education, our daily decisions, our politics, even our sports and pastimes. It lies at the heart of our purest intellectual activities and our unparalleled ability to control the world around us. If one were to list the most distinctive and important of human activities, science would surely be among them.

In view of this significance, it should be unsurprising that science *itself* has become an object of intensive enquiry. Thus, researchers from a broad array of different disciplines—including history, sociology, economics, psychology, and philosophy—have sought to understand various aspects of science. Further, over the past few decades, newly emerging interdisciplinary fields, such as *science and technology studies* (STS), have sought to integrate the deliverances of such research in pursuit of a more complete understanding of science.

Though a detailed discussion of these various fields of research falls well beyond the scope of the present chapter, we do need to say more about the philosophy of science and its potential interconnections with the cognitive and behavioral sciences—especially psychology. Psychologists have long been interested in the sorts of cognitive activities central to science—explanation, reasoning, inductive learning, categorization, and concept formation, for example. Moreover, when psychologists study such phenomena, they typically deploy a standard battery of experimental methods. Similarly, philosophers of science have long been interested in questions about the nature of science, its practices, and its core concepts. And when philosophers of science address such issues, they too draw on their own battery of familiar methods—including, for example, the construction of arguments and counterexamples, the analysis of concepts, and, on occasion, the use of logics and formal models.

Strikingly—and this is the departure point for the present volume—there is remarkably little systematic interaction between the philosophy of science and the sorts of experimental approaches to be found in psychology. A core assumption of the present volume is that this lack of interaction presents an intriguing opportunity for growth. Philosophical and psychological approaches to the study of science *should* interact in deeper, more systematic ways than they currently do. This volume is thus dedicated to exploring the prospects for an *experimental* philosophy of science—one that uses the empirical methods of psychology in order to help address questions in the philosophy of science.

Experimental philosophy and philosophy of science

The rarity with which experimental methods from psychology have been applied by philosophers of science is puzzling, given the recent impact of such methods on many other areas of philosophy. Epistemologists, for example, have become increasingly interested in experimentally investigating how such concepts as *knowledge* and *justification* are deployed in people's explicit judgments (see, e.g., Beebe, 2014). Circumstances are similar in the philosophy of action, metaethics, the philosophy of language, the philosophy of mind, and even some parts of metaphysics. In each of these fields, one finds a growing group of philosophers using experimental psychological techniques to help develop and assess accounts of philosophically significant concepts, such as *reference*, *consciousness*, and *intention*. (For recent reviews of such work, see Sytsma and Buckwalter, 2016.) Appropriately enough, the resulting approach has been dubbed *experimental philosophy* (Knobe et al., 2012).

Though there are many reasons for this "experimental turn," one highly influential consideration, widely associated with the so-called *positive program*, is to aid in what has long been a core philosophical activity—the analysis of philosophically important concepts (Knobe and Nichols, 2008). Ordinarily, when philosophers engage in conceptual analysis, they draw heavily on their own "intuitions" in order to assess proposals regarding the structure and content of the salient concepts. Experimental philosophers maintain that the use of empirical methods—applied to broad, (ideally) representative populations—can make a substantial contribution to this project. In particular, they maintain that, by adopting this approach,

> [t]hey can avoid some of the idiosyncrasies, biases, and performance errors that are likely to confront philosophers who attend only to their own intuitions and the intuitions of a few professional colleagues who read the same journals and who may have prior commitments to theories about the concepts under analysis. (Stich and Tobia, 2016, p. 23)

So construed, experimental philosophy might be viewed as an attempt to replace potentially idiosyncratic judgments with more thorough empirical research. What's novel is that experimental philosophers propose that scientific standards ought to apply to philosophy as well. More specifically, they maintain that, to the extent that philosophical theories incur empirical commitments regarding people's intuitions and other psychological states, philosophers should aspire to the same standards as scientists do.

Given the careful attention that philosophers of science have paid to experimental methodology, one might have expected to find them at the forefront of this new development in philosophy. However, for reasons that are not immediately obvious (at least not to us), philosophers of science have been slow to employ experimental methods of any sort.[1] Moreover, what work has been done is isolated and scattered. Thus, the primary goal of this volume is to bring together research that both explores and exemplifies the prospects for an experimental philosophy of science.

What might experimental philosophy of science be?

As we see it, the core idea behind experimental philosophy of science is to use empirical methods, typically drawn from the psychological sciences, to help investigate questions of the sort associated with the philosophy of science. This conception is, by design, ecumenical in a variety of ways.

First, in our view, experimental philosophy of science is to be characterized in terms of subject matter and methodology and not in terms of *who* is doing the research. To be sure, there are "card-carrying" philosophers who engage in the sorts of research we have in mind. But it is at least as common to find professional scientists applying the methods of psychology to questions associated with the philosophy of science. In our view, these scientists are as much a part of experimental philosophy of science as any philosopher. Indeed, in view of the extensive experimental training that such scientists possess—training seldom possessed by professional philosophers—there are obvious reasons why, when viable, philosophers engaged in experimental work should collaborate with behavioral scientists. This attitude is very much reflected in the coming chapters, many of which are (co)authored by behavioral scientists as well as self-identified philosophers.

Second, our characterization of experimental philosophy of science imposes no explicit restrictions on which empirical *methods* might be relevant to addressing issues in the philosophy of science. The techniques of psychological science are quite extensive, and we see no *a priori* reason to exclude the possibility that relevant results might come from any of a range of quite different sorts of research, including developmental research, reaction-time studies, patient studies, and functional Magnetic Resonance Imaging (fMRI) research. As a matter of fact, however, most extant research in experimental philosophy— including some of what's reported in this book—relies on a kind of protocol in which stimuli are presented in written form—as "vignettes"—and responses are elicited via probe questions and recorded on Likert-type scales. Such studies are sometimes disparagingly referred to as "surveys." But this is unfair. It's true that they are survey-*like* in that both stimuli and probes are presented linguistically, and explicit judgments are elicited. However, it's important to keep in mind that, in contrast to surveys—which merely seek to record people's views— research in experimental philosophy almost invariably involves the control and manipulation of different variables. No doubt that there are interesting methodological issues regarding the scope and limits of such techniques,[2] but they are not as easily disparaged at the "survey" label might suggest. Indeed, these methods are commonplace in many areas of psychology—including the judgment and decision-making literature—and possess some notable pragmatic virtues. Most obviously, they are inexpensive, relatively simple to design and administer, and they deliver results that are relatively easy to analyze.

Third, we are inclined to adopt an ecumenical conception of the sorts of issues that fall within the purview of philosophy of science. This is, in large measure because, philosophers of science *themselves* seem to adopt such a view. To be sure, there's the list of familiar "canonical" questions that one might seek to cover in a survey of general philosophy of science—for example, questions about the demarcation of science from nonscience and pseudoscience; issues about our concept of explanation; issues about causation and our concepts thereof; issues about what reduction is; issues about theory change in science; and issues about the rationality of science. But there are also issues that are rather less well-worn—for example, about the role of moral and political values in science, and about the epistemic status of thought experiments. Moreover, when one turns to the various philosophies of specific sciences—of biology, physics, economics, psychology and chemistry, for example—one finds a fascinating array of issues concerning the methods, practices, and concepts of these different scientific fields. In the philosophy of biology, for example, we find issues regarding the notion of *function* and its role in biological science, issues about the concept(s) of a *gene*, and issues about the extent to which explanation in biology is mechanistic. Some of these topics are taken up in the chapters of this volume. But in our view, there are many more issues that might form a focus for experimental philosophy of science. Given the immaturity of the field, we'll have to wait and see.

Finally, we remain largely neutral regarding the precise *extent* to which experimental psychological research might contribute to addressing issues in the philosophy of science. At one extreme, there is a vision of experimental philosophy we've encountered in conversation, which conceives its goal as being the wholesale replacement of traditional philosophical research by a discipline in which questions are almost exclusively addressed by empirical means. This is not a result we find either realistic or desirable. Our goal is rather less colonial. It is simply that experimental techniques should become a commonly used addition to the already expansive range of methods that can be brought to bear on issues in the philosophy of science.

Why *experimental* philosophy of science?

Philosophy has long been rife with methodological debate, and in recent years, few such debates have been more heated than those regarding the value of experimental philosophy. We don't propose to rehearse these general metaphilosophical issues here. Instead, we make a few quite specific suggestions regarding why, and how, experimental philosophy of science might be a valuable endeavor.

Experimental philosophy of science as an extension of STS

Here's a preliminary consideration: one reason to engage in experimental philosophy of science is as a natural extension of STS more generally. Science is an extremely important human endeavor, as worthy of study as almost any other institution or social arrangement. Many cognitive and behavioral scientists are interested in various questions pertaining to how science is practiced. Experimental philosophy of science would be quite similar in its aims and methods but would be primarily concerned with, and motivated by, questions more typical of the philosophy of science. For example, philosophers of science are interested in questions regarding the nature of explanation, causation, and understanding, as well as the nature of important scientific constructs, such as theories and models. To the extent that such questions are worthy of pursuit, empirically minded investigations are one important way of doing so. On such a view, then, experimental philosophy of science would simply be one subdomain of STS—a kind of "cusp" point where the psychology and the philosophy of science intersect.

Experimental philosophy of science as an extension of the "turn to practice"

One might, however, wonder about the specific *philosophical* significance of experimental philosophy of science. Why, in particular, should we think it worthwhile, in the first place, to apply experimental methods in addressing philosophical questions about science?

As a rule, we are wary of such questions since they often assume a relatively sharp divide between what's philosophical and what's not. In the present context, however, we think there's a rather more local point to be made—one that's quite specific to the current state of the philosophy of science and concerns the so-called "turn to practice" (Soler et al., 2014, p. 1).

The story behind the "turn to practice" is a familiar one. By the 1970s, traditional philosophy of science was deemed by many "analysts of science" to be too idealized—"too disconnected from how science actually is performed in laboratories and other research settings" (Soler et al., 2014, p. 1). In particular, ethnographic studies conducted in scientific laboratories showed that science was a complex activity that involved much more than experiments and logical inferences (Soler et al., 2014, p. 1).

As a consequence, both philosophers and social scientists increasingly paid greater "attention to scientific practices in meticulous detail and along multiple dimensions, including the material, tacit, and psycho-social ones" (Soler et al., 2014, p. 1).

We join the consensus in applauding the turn to practice. Not only does it help debunk excessively idealized conceptions of science, but also it imposes a plausible, though defeasible, condition of adequacy on philosophical accounts of science. Roughly put: All else being equal, a philosophical account of some aspect of science—explanation, reduction, genetics, etc.—ought not to be inconsistent with the extant practices of the relevant group of scientists.

It is worth noting, however, that the methods which originally motivated the turn to practice were largely drawn from the social sciences—especially sociology and anthropology—and not from psychology. But if the turn to practice is concerned with what actual scientists *do*, then it is surely the case that *cognitive* activities—thinking, reasoning, learning, and the like are things that scientists do *qua* scientists. Arguably, they are among the core practices of science. And if this is so, then we seem to have good reason to suppose that the methods of psychology are relevant to the philosophy of science. After all, it is very plausible that they are our best methods for studying cognitive activity.

On the potential contributions of experimental philosophy of science

So far, we have argued that experimental psychological methods ought to be taken seriously when addressing questions in the philosophy of science. Yet, it is one thing to say they should be taken seriously, but quite another to say what they would, in fact, contribute to the philosophy of science. Plausibly, this will depend on the sort of question that's at issue, and there may well be many contributions that experimental methods might make. For the moment, however, we mention just four:

1. Conceptual diversity. As noted above, philosophers of science have long sought to characterize concepts that figure prominently in one or other region of the sciences. As Paul Griffiths and Karola Stotz note, however, a tacit assumption of much of this research is that the philosopher is sufficiently well informed about the relevant field to be "in a position to consult his or her intuitions as a scientifically literate sample of one, and thus equivalent for this purpose to a member of the scientific community" (Griffiths and Stotz, 2008, p. 1). (Compare: A linguist might, as competent speaker of their *own* dialect, consult intuitions about the well-formedness of sentences, thereby avoiding the need to consult other speakers.) However, this approach faces obvious difficulties if the "key scientific concepts display substantial heterogeneity between different communities of researchers" (Griffiths and Stotz, 2008, p. 1). Under such circumstances, the intuitions of a single respondent obviously won't reveal conceptual diversity *between* communities, any more than the linguist's intuitions regarding their *own* dialect will reveal variations between different dialects. At the risk of stating the obvious, experimental methods provide a natural approach to acquiring data about cross-community variation.

2. Conceptual uniformity. The use of experimental methods in philosophy of science is not merely motivated by an interest in conceptual diversity; it may also be motivated by its converse: the search for general philosophical proposals that apply generally across both scientific and nonscientific contexts. Consider the case of explanation. As Woodward (2017) observes, there is a widespread tendency in the recent philosophical literature "to assume that there is a substantial continuity between the sorts of explanations found in science and at least some forms of explanation found in more ordinary nonscientific contexts." Though the motivations for this assumption are not entirely obvious, there are two *prima facie* reasons that readily come to mind. First, in contrast to such concepts as a *Higgs boson* or *lateral geniculate nucleus*, which are products of science and make little sense independently of this context, the concept of explanation really seems to have led a life outside the lab. Second, in view of this, the assumption of substantial continuity seems like the reasonable default position to adopt. All else being equal, more unified accounts are preferable to less unified ones. Yet if we seek such unity, then we have a good reason to empirically study the explanatory practices and judgments of different populations, both within and outside of science. For, by doing so, we can identify those features of explanation that are highly conserved across disparate populations, and also assess the assumption of continuity which motivates the endeavor in the first place.

3. Modeling scientists' effective concepts. A third reason to deploy experimental methods in the philosophy of science is that it promises to provide us with a deeper grasp of the concepts that scientists use. When philosophers attempt to explicate scientific concepts, they often rely on textbooks or the writings of important scientists in the relevant disciplines. But as Eduoard Machery has pointed out, it is quite possible that the *explicit* concepts found in textbooks or influential writings differ from scientists' *operative* concepts—the ones they in fact deploy in their daily, professional activities. Among other things, this might be so because scientists may be unusually reflective in such writings or because concept use in textbooks or influential writings may lag behind concept use in the research front (Machery, 2016, p. 476).

In view of this, if one seeks to characterize the operative concepts that scientists deploy in actual practice, then the sorts of experimental methods found in psychology would, once more, appear relevant.

4. Cognitive foundations. A final, related role for experimental philosophy concerns its potential to contribute to an understanding of the cognitive foundations of science. Philosophers of science have long had interests in questions concerning those cognitive processes most distinctive of science—for example, conceptual change, theory formation, inductive learning, and causal inference (Hempel, 1952). Until recently, however, there has been a widespread tendency to "depsychologize" such topics, either by treating them as normative ones, or else by adopting a level of idealization which abstracts from almost any empirical content regarding how human beings in fact engage in such activities. Yet if one seeks to capture actual scientific practice, more realistic models are required; and once again, this provides clear motivation for deploying the methods of psychology, and cognitive science more broadly (Carruthers, Stich and Siegal, 2002; Nersessian, 2008; Knobe and Samuels, 2013).

Forthcoming attractions

Though the chapters in this volume illustrate a wide array of different concerns, we start with one of the most central and enduring topics in the philosophy of science. At least since Hempel and Oppenheim (1948), philosophers of science have been acutely aware of the need for an account of explanation in order to understand scientific practice. However, only relatively recently has there been any serious effort to empirically examine the role that explaining plays in our cognitive lives. In Chapter 2, Elizabeth Kon and Tania Lombrozo explore this issue by focusing on the way that efforts at explanation influence people's ability to identify generalizations. In particular, they focus on the differing roles played by principled, as opposed to seemingly arbitrary, exceptions to generalizations.

Continuing the work on explanation, in Chapter 3, Frank Keil addresses one of the mysteries surrounding children's and adults' preference for mechanistic explanations. The mystery is that people seem to be rather bad at *remembering* how mechanisms operate (Rozenblit and Keil, 2002; Mills and Keil, 2004.) But if people tend to forget how mechanisms operate, then why do both children and adults prefer mechanistic explanations to the alternatives? In response to this puzzle, Keil argues that while individuals forget the details of mechanistic explanations, exposure to such explanations provides access to higher-order causal patterns that prove invaluable for various other purposes. Indeed, he suggests that access to such patterns explains another mystery: why both children and adults fail to recognize the deficits in their own mechanistic understanding.

Another longstanding issue in philosophy of science concerns how people go about changing their minds in light of new evidence. Traditionally, most philosophers 70+ years ago would have had it that people change their minds in light of new evidence. However, Thomas Kuhn (1962/2012) famously argued that at least in some circumstances, theory change was not based on rational considerations. In a more modern variant, Kahan et al. (2012) have argued that, at least with respect to some domains, people exhibit a remarkable degree of insensitivity to new scientific information. In particular, they argue that people's attitudes toward politically charged scientific questions, such as the existence and importance of anthropogenic climate change, remain almost entirely impervious to new information. Fighting against this tide, Ranney et al. (e.g., Ranney & Clark 2016) have argued that, even on such matters, people's attitudes are in fact far more sensitive to relevant information than Kahan et al. would have us believe. In Chapter 4, Michael Ranney, Matthew Shonman, Kyle Fricke, Lee Nevo Lamprey, and Paras Kumar investigate a surprising new way to change people's minds about scientific truths—artificially inflating or decreasing their nationalism. This suggestion has immediate practical implications regarding how to increase science literacy. However, whether the overall moral is a positive one (people have coherent worldviews that are sensitive to evidence) or a negative one (people can be manipulated in all sorts of indirect ways) remains something of an open question.

Addressing another topic regarding theory change, philosophers have questioned what prevents people from developing and accepting new and better theories. Scientists often have difficulties producing and understanding new theories, as of course do students. One might have thought that these difficulties were congruent—the same conceptual barriers to scientists' developing and accepting new theories also impede the progress of students. In Chapter 5, Andrew Shtulman investigates the data on what sorts of mistakes students are likely to make in order to argue that they are frequently stymied by very different sorts of problems from those that slowed down the development of science itself. This has implications for both science education and the study of theory change.

Focusing on knowledge acquisition in children, in Chapter 6, Mark Fedyk, Tamar Kushnir, and Fei Xu argue that the only way to make sense of the barrage of data imposed on us by the world is to already have certain intuitive concepts that allow children to find the properties relevant to accurate belief formation. More specifically, they argue that by the age of four, children have a *theory of evidence* that enables them to make sense of the world in a manner conducive to accurate belief formation.

Turning from general questions in philosophy of science to issues more particular to individual sciences, in Chapter 7, Michiru Nagatsu takes a broad look at the potential of experimental philosophy to address issues in the philosophy of economics. Specifically, he looks at different notions of "choice," "preferences," and "nudges"—examining both how such constructs are understood in economics and among the public, and how experimental philosophy can provide novel insight.

In the philosophy of biology, scientists have recently begun worrying that the concept of *innateness* is an amalgam of different ideas (fixity, typicality, and functionality) which—if taken seriously—seem to license invalid inferences from the presence of one of these features to the others. While one might not worry about the folkbiological use of innateness, its continued usage by scientists is somewhat puzzling. In Chapter 8, Edouard Machery, Paul Griffiths, Stefan Linquist, and Karola Stotz explore the question of whether scientists are really using the term to pick out a new and better-behaved concept, or whether in practice they are falling back on the old and seemingly broken version. They find evidence for the latter hypothesis, indicating that the folkbiological notion of innateness is alive and well among the behavior of practicing scientists.

Finally, in Chapter 9, we explore a view of the relevance of experimental results for the philosophy of causation. James Woodward argues that philosophy of science is at its best a normative enterprise, and so some uses of empirical data regarding how a concept like causation *is* used leave open the more pressing question of how it *should* be used. While this could be taken as a pessimistic interpretation of the role of experimental philosophy generally, it also suggests a path forward for how experimental philosophy can be deployed in a way that provides genuine guidance regarding key issues in philosophy of science. This seems a fitting note on which to end, for suggesting future avenues of exploration is in large measure the goal of the present volume.

Notes

1 This is so despite early persuasive advocacy by Griffiths and Stotz (2008).
2 For more extensive discussion, see Chapter 9 of this volume.

References

Beebe, J. (ed.) (2014). *Advances in Experimental Epistemology*. New York: Bloomsbury Academic.

Carruthers, P., Stich, S. and Siegal, M. (eds.) (2002). *The Cognitive Basis of Science*. New York: Cambridge University Press.

Griffiths, Paul E. and Stotz, K. (2008). Experimental philosophy of science. *Philosophy Compass*, 3(3): 507–21.

Hempel, C. G. (1952). *Fundamentals of Concept Formation in Empirical Science*. Chicago: University of Chicago Press.

Hempel, C. G. and Oppenheim, P. (1948). Studies in the logic of explanation. *Philosophy of Science*, 15(2): 135–75.

Kahan, D. M., Peters, E., Wittlin, M., Slovic, P., Ouellette, L. L., Braman, D., and Mandel, G. (2012). The polarizing impact of science literacy and numeracy on perceived climate change risks. *Nature Climate Change*, 2(10): 732.

Knobe, J., Buckwalter, W., Nichols, S., Robbins, P, Sarkissian, H., and Sommers, T. (2012). Experimental philosophy. *Annual Review of Psychology*, 63(1): 81–99.

Knobe, J. and Nichols, S. (2008). An experimental philosophy manifesto. In J. Knobe and S. Nichols (Eds.), *Experimental Philosophy* (vol. 1, pp. 3–14). New York: Oxford University Press.

Knobe, J. and Samuels, R. (2013). Thinking like a scientist: Innateness as a case study. *Cognition*, 128: 72–86.

Kuhn, T. S. (1962/2012). *The Structure of Scientific Revolutions*. Chicago: University of Chicago Press.

Machery, E. (2016). Experimental philosophy of science. In S. Sytsma and W. Buckwalter (Eds.), *A Companion to Experimental Philosophy* (pp. 475–90). Malden, MA: Wiley-Blackwell.

Mills, C. M. and Keil, F. C. (2004). Knowing the limits of one's understanding: The development of an awareness of an illusion of explanatory depth. *Journal of Experimental Child Psychology*, 87(1): 1–32.

Nersessian, N. (2008). *Creating Scientific Concepts*. Cambridge: MIT Press.

Ranney, M. A. and Clark, D. (2016). Climate change conceptual change: Scientific information can transform attitudes. *Topics in Cognitive Science*, 8: 49–75. Doi: 10.1111/tops.12187

Rozenblit, L. and Keil, F. (2002). The misunderstood limits of folk science: An illusion of explanatory depth. *Cognitive Science*, 26(5): 521–62.

Soler, L., Zwart, S., Lynch, M., & Israel-Jost, V. (2014). *Science after the practice turn in the philosophy, history, and social studies of science*. Routledge.

Stich, S. and Tobia, K. (2016). Experimental philosophy and the philosophical tradition. In J. Sytsma and W. Buckwalter (Eds.), *A Companion to Experimental Philosophy* (pp. 3–21). Malden, MA: Wiley Blackwell.

Sytsma, S. and Buckwalter, W. (Eds.) (2016). *A Companion to Experimental Philosophy*. Malden, MA: Wiley-Blackwell.

Woodward, J. (2017). Scientific Explanation. In E. N. Zalta (Ed.), *Stanford Encyclopedia of Philosophy*. Stanford, CA: The Metaphysics Research Lab.

Part One

Explanation and Understanding

2

Scientific Discovery and the Human Drive to Explain

Elizabeth Kon and Tania Lombrozo

Carl Hempel suggested that two human concerns provide the basic motivation for all scientific research (Hempel, 1962). The first is "man's persistent desire to improve his strategic position in the world by means of dependable methods for predicting and, whenever possible, controlling the events that occur in it." The second is "man's insatiable intellectual curiosity, his deep concern to *know* the world he lives in, and to *explain*, and thus to *understand*, the unending flow of phenomena it presents to him." Hempel isn't alone in highlighting a special role for explanations in science: others identify explanatory theories as the "crown of science" (Harre, 1985), with explanations as the "real payoff" from doing science (Pitt, 1988).

Why are explanations at the heart of science and scientific advance? In this chapter, we propose that explanations play a crucial role in scientific discovery, thereby advancing Hempel's first motivation for scientific research: the achievement of a better strategic position in the world through better prediction and control. The value of explanation is thus in large part instrumental (Lombrozo, 2011), with the quest for explanations driving scientific theory construction, and the generation of explanations linking theory to application.

The motivation for our proposal comes from recent work in cognitive psychology on the role of explanation in learning. This work suggests that the very process of seeking explanations motivates children and adults to go beyond the obvious in search of broad and simple patterns, thereby facilitating the discovery of such patterns, at least under some conditions (Lombrozo, 2016). Might the drive for scientific explanation play a similar role in science, prompting individuals and communities to search deeper and harder for broad and simple generalizations that characterize the natural world? Could features of explanatory cognition themselves explain features of scientific practice and theorizing, such as the allure of exceptionless laws and simple theories?

To a large extent, these questions remain unanswered: there has been little empirical research on the role of explanation in actual scientific practice,[1] nor will we report such research here. Instead, our aim is to evaluate how research on everyday human cognition might extend to scientific contexts by reviewing prior psychological research, and by presenting new studies that explore the effects of explanation in a learning environment that is more representative of most scientific practice. Specifically, we explore a puzzle that arises from prior research. On the one hand, this research suggests that when people engage in explanation, they aim to achieve an explanatory ideal: obtaining explanations that are underwritten by *simple and exceptionless* patterns or generalizations. On the other hand, we know that in real scientific practice, such generalizations are rarely to be found. Could it be that the search for ideal explanations is beneficial in part because it facilitates the discovery of real but imperfect generalizations— for example, those that involve some exceptions? (For impatient readers, we offer a hint: the answer is "yes.") But before turning to our new studies, we briefly review relevant prior work.

The role of explanation in learning

Decades of research reveal that the process of explaining—even to oneself—can have a powerful effect on learning (e.g., Fonseca and Chi, 2011; Lombrozo, 2012; Chi et al., 1989). Several psychological processes contribute to this phenomenon. For example, attempting to explain something can help people appreciate what they do not know (Rozenblit and Keil, 2002), make them accommodate new information within the context of their prior beliefs (Chi et al., 1989; Williams and Lombrozo, 2013), and lead them to draw inferences to fill gaps in their knowledge (Chi, 2000). There is also evidence that when engaged in explanation, both children and adults seek explanations that are *satisfying*, where satisfying explanations are those that account for what is being explained by appealing to broad and simple rules or patterns (Lombrozo, 2016). For example, Williams and Lombrozo (2010, 2013) found that when presented with an array of items belonging to two categories, adults who were prompted to explain why each item belonged to its particular category (e.g., why robot A is a "glorp" and robot B is a "drent") were more likely than those in control conditions to discover a subtle classification rule that accounted for the category membership of all items on the basis of a single feature (see also Walker, Bonawitz, and Lombrozo, 2017). This was true whether participants in the control condition were prompted to describe the category exemplars, to think aloud as they studied them, or to simply engage in free study.

If explanation is so beneficial for learning, one might wonder why people don't explain more often. In other words, why don't children and adults engage in explanation spontaneously, even in the "control" conditions that are used as a baseline against which to compare the performance of participants who are explicitly prompted to explain? To some extent, people do explain without an explicit prompt: participants explain to varying degrees, even in control conditions, and this variation predicts what they ultimately learn (Edwards et al., 2019; Legare and Lombrozo, 2014). But there's more to the story than that. It's possible that people are frugal explainers not only because it is effortful to explain, but also because *explicitly engaging in explanation does not always yield superior performance*. Under some conditions, prompts to explain result in learning that is no different from that in a control condition (Kon and Lombrozo, in prep), suggesting that explanation is unnecessary. Under other conditions, prompts to explain can actually be detrimental (Williams, Lombrozo, and Rehder, 2013; see also Legare and Lombrozo, 2014; Walker et al., 2014). It's instructive to consider these cases in turn.

First, Kon and Lombrozo (in prep) identify conditions under which a prompt to explain is unnecessary in the sense that it does not lead to performance that exceeds that of control conditions. They find that when it comes to discovering a subtle, exceptionless pattern describing a set of observations, participants who are prompted to explain only surpass those in a control condition when there is a compelling but inferior pattern for those in the control condition to latch on to, such as a salient pattern that accounts for 75 percent of observed cases. For nonexplainers, this alternative is sufficiently compelling to limit the further expenditure of cognitive resources. But for explainers, a pattern that only accounts for a subset of cases, or that does so in a complicated way, isn't good enough; the expectation or hope of a more satisfying explanation spurs them on. Studies with young children similarly hint at the idea that explaining is a spur to go "beyond the obvious" to find a pattern that is more subtle, but also in some regards superior, to more salient possibilities (Walker et al., 2014; Walker, Bonawitz, and Lombrozo, 2017). In the case of science, this could mean that seeking explanations (and not, say, mere descriptions) is likely to spur additional discoveries when the discoveries go beyond salient regularities to capture generalizations over nonobvious properties.

Second, Williams, Lombrozo, and Rehder (2013) find that under some conditions, a prompt to explain observations can actually be detrimental. In their task, participants had to learn how to classify vehicles into two categories, or how to identify individuals as likely or unlikely to make charitable donations. In some cases, these tasks could be achieved by identifying a single theme or feature that characterized all members of one category and none of those in the other. But in other conditions, the only way to achieve perfect classification was to memorize the idiosyncratic properties of individual exemplars—for instance, that the cyan car was a "kez" or that *Janet* frequently donates to charities. Under these conditions, those participants who were prompted to explain took longer to learn how to accurately categorize all of the exemplars; they seemed to perseverate in looking for a broad and simple *pattern* before settling for a rote strategy based on individuals. Generalizing to science, we might expect the search for explanations to be detrimental when there is *no structure at all* to the observations being explained. This is probably an unusual situation for science, but it might arise when a set of observations is grouped according to a wholly inaccurate theory or when observations reflect noise rather than some underlying signal.

What is it about broad and simple patterns that satisfies the demands of explanation? Or conversely, what is it about patterns with exceptions or additional complexity that *fails* to satisfy the demands of explanation? Recent work by Kon and Lombrozo (in prep) contrasts two possibilities: that explainers favor exceptionless patterns because such patterns maximize predictive power, or that explainers favor exceptionless patterns because such patterns make for more virtuous explanations—that is, explanations that exhibit the explanatory virtues of simplicity and breadth. To differentiate these alternatives, they created learning tasks in which participants could achieve perfect predictive accuracy on the basis of two salient features of the stimuli (thus achieving breadth at the expense of simplicity), or potentially discover a more subtle pattern that also supported perfect predictive accuracy and did so on the basis of a single feature (thus achieving both breadth and simplicity, but at a cost of greater cognitive effort). Participants who were prompted to explain were significantly more likely than those in a control condition to discover the more subtle rule. This suggests that the salient, predictively perfect (but less virtuous) alternative was insufficient to satisfy their explanatory drive. This fits well with a familiar observation from science: the most predictive model isn't always the most explanatory. Explanation seems to require something more than successful prediction.

To sum up, prior work on the effects of explanation on learning suggests that when people are actively engaged in seeking explanations for particular observations, they're more likely to find simple, exceptionless patterns that underlie those observations, and that these patterns are compelling because they support virtuous (i.e., simple and broad) explanations. Explanation is not always beneficial (relative to control conditions), but it does appear to be beneficial when two conditions obtain: when there is a broad and simple pattern to be found, and when there is a more salient but inferior alternative for participants in the control condition to latch on to. It may not be accidental, then, that science focuses so heavily on explanation. The natural world does appear to be bursting with patterns, many of which can only be formulated over unobservable and otherwise nonobvious properties. Nature seems to reward those who not only consider the obvious but also go beyond it. These findings also resonate with a feature of how "idealized" physics is often portrayed: as a search for exceptionless laws and elegant theories. Even in domains that don't aspire to exceptionless laws, there seems to be value in minimizing and accounting for exceptions. The search for simple and broad generalizations thus seems to act as a powerful motivating force: it's not enough to find a pattern; it must be a pattern of the right sort.

Despite these synergies between our experimental studies and observations about science, an important puzzle remains. After all, scientists rarely succeed in identifying truly exceptionless laws. Especially within the social sciences, generalizations are invariably imperfect and riddled with exceptions. In some domains, accounting for even 75 percent of the variance in the manifestation of some property (such as personality) is a notable achievement. Could it be that engaging in explanation motivates everyday learners—and scientists—to search for simple, exceptionless patterns, but that in the course of doing so, *they're also more likely to discover other subtle but imperfect regularities that nonetheless constitute an advance*?

Evidence that this could be so comes from Experiment 3 of Kon and Lombrozo (in prep), in which participants were tasked with learning how to determine whether novel creatures eat flies or eat crabs. Half the participants were prompted to write down an explanation for each observation (i.e., for why a particular creature eats flies or crabs), and half (in the control condition) were prompted to write down their thoughts about that observation. The observations were designed to support two possible generalizations. First, participants could learn to predict the diet of all studied examples on the basis of two features of the stimuli, their habitat *and* age, which was a complex but exceptionless pattern. Second, participants could learn to predict the diet of a majority of studied examples (75 percent) on the basis of a single feature—snout direction—which was a simple rule, but one with exceptions. Kon and Lombrozo found that participants who were prompted to explain were more likely than those in the control condition to discover each of these rules, presumably because they stumbled across them in their search for an ideal explanation: one that was *both* simple and exceptionless. This finding suggests that even if a simple, exceptionless pattern describes some explanatory ideal that is rarely realized, the pursuit of this ideal could spur meaningful discoveries.

So far we've considered the role of explanation in learning, and how prior work on explanation might shed light on the puzzle of whether and why seeking ideal explanations is beneficial, given that we inhabit a less-than-ideal world. In what follows, we describe a pair of novel experiments designed to test our core hypothesis in a more systematic fashion: that in pursuing an ideal explanation, explainers increase their odds of discovering some of the nonideal structure to be found. In Experiment 1, we thus present learners with a nonideal world (i.e., one that does not support a maximally simple and broad generalization) and investigate the effects of a prompt to explain.

Explaining in a nonideal world: Two novel experiments

Experiment 1: Is explaining beneficial when all generalizations involve exceptions?

Experiment 1 investigates whether in the absence of an ideal pattern (i.e., one that is both maximally simple and broad), engaging in explanation can nonetheless assist with the discovery of the best available alternatives. To test this, we designed a task in which participants learned to categorize items into one of two categories. As they studied twelve labeled exemplars (six from each category), they were prompted either to explain or to write down their thoughts about the category membership of the exemplars. Two rules could be used to categorize the items. One rule was fairly salient and therefore easy to discover, but only captured the category membership of eight of the twelve exemplars (it was thus a "66 percent rule"). Another rule was much subtler but captured the category membership of ten of the twelve exemplars (it was thus an "83 percent rule"). So while the latter rule still fell short of the ideal (i.e., a rule that captured all twelve items, a "100 percent rule"), it was superior to the initial rule along the dimension of breadth.

If explaining assists in the discovery of the best possible rule, even if it is imperfect, we would expect participants prompted to explain to be more likely than those in the control condition to discover the 83 percent rule. By contrast, if effects of explanation are restricted to the ideal case—an exceptionless rule[2]—then we would expect those participants who were prompted to explain to perform no better than those in the control condition.

In addition to the *no ideal rule* condition just described, we also considered an *ideal rule* condition, in which the more salient rule accounted for 83 percent of cases, and the more subtle rule accounted for 100 percent of cases. This more familiar situation is a replication of prior research, but with a larger number of training exemplars (twelve versus eight) to accommodate the intermediate percentages. We included it in Experiment 1 in part as an extension of prior research but also to serve as a basis for comparison against the *no ideal rule* condition. Thus we can ask not only whether a prompt to explain facilitates discovery of a "better" rule when the better rule is an 83 percent rule (versus a 66 percent worse rule), but also whether the magnitude of this effect is comparable to the effects of explanation when the "better" rule is a 100 percent rule (versus an 83 percent worse rule).

Method

Participants

The sample for Experiment 1 consisted of 1293 adults[3] (after exclusions)[4] recruited through Amazon Mechanical Turk and paid for their participation. In all studies, participation was restricted to adults with an IP address within the United States and with an approval rating of at least 95 percent on fifty or more previous tasks. Participants were also prevented from participating in more than one study from this paper.

Materials

The stimuli consisted of ten sets of twelve items. The twelve items in each set depicted flowers, containers, objects, simple robots, or complex robots. Throughout this chapter, we will use flowers as an illustrative example, but information concerning the other four stimulus types is available in Appendix A.

Each set contained items from two categories, with six items belonging to each category. For example, half of the flower items were SOMP flowers and half of the flower items were THONT flowers. For each set, participants could use two possible rules to determine which category an item belonged to. One rule was always "better" in the sense that it could be used to correctly categorize more items than the "worse" rule. In the *ideal rule* condition, the better rule was a "100 percent rule" that perfectly accounted for the category membership of all twelve items, and the worse rule was an "83 percent rule" that correctly categorized only ten of the twelve items (see Figure 2.1). For the *no ideal rule* condition, the better rule was an "83 percent rule" that correctly categorized ten of the twelve items, and the worse rule was a "66 percent rule" that correctly categorized eight of the twelve items.

Procedure

The task consisted of a study phase followed by a reporting phase and a rule rating phase. At the start of the study phase, participants were randomly assigned to one of four conditions, which were created by crossing two prompt-types, *Explain* or *Write Thoughts*, with two pattern-types, *ideal rule* or *no ideal rule*. Participants were randomly assigned to see one of the stimulus sets.[5]

In the study phase, all participants were told to study the items, and that after the study phase they would be asked questions about how to determine which category each item belongs to. Participants were presented with a randomized array of the twelve items corresponding to their condition's pattern-type

Figure 2.1 Flower stimuli. For these flower stimuli, the better rule (100 percent in the *ideal rule* condition and 83 percent in the *no ideal rule* condition) is that SOMP flowers have two concentric circles in their centers, whereas THONT flowers have one circle in their centers, and the worse rule (83 percent in the *ideal rule* condition and 66 percent in the *no ideal rule* condition) is that the petals of SOMP flowers are mostly one color (in the colored version, these are all different colors; in grayscale, they are all the same tone to increase the salience of the worse rule), while the petals of THONT flowers are mostly rainbow-colored (indicated in grayscale with higher-contrast leaves). Within this figure, the double solid outline contains the items in the *no ideal rule* condition, and the double dotted outline contains the items in the *ideal rule* condition. The exceptions to the worse rule are in solid boxes and the exceptions to the better rule are in dashed boxes.

(*ideal rule* or *no ideal rule*). They were then prompted to focus their attention on each item, individually, in a random order, with a prompt determined by the experimental condition to which they were randomly assigned. Participants in the *explain* conditions were told (for example) to "try to *explain why* flower A is a SOMP flower." Participants in the *write thoughts* conditions were told to "*Write out your thoughts* as you learn to categorize flower A as a SOMP flower." Participants were given 50 seconds to respond to each prompt by typing into a text box, at which time their responses were recorded and the prompt for the next item appeared.

In the reporting phase, participants were asked to report all patterns that they noticed that differentiated SOMPS and THONTS, even if the patterns were imperfect.[6] In addition to describing the rule they discovered in a free-response box, participants were asked how many of the twelve items they thought followed the rule.

After finishing the reporting phase, participants were again presented with all twelve items as well as four candidate rules, presented in a random order, purporting to explain "why flowers A–F are SOMPS (as opposed to THONTS)." They were forced to stay on the page for at least 15 seconds to ensure that they read the explanations. The candidate rules referenced the better rule, two versions of the worse rule (one indicating that it involved exceptions, one not), and one filler item that was a bad/untrue explanation (see Appendix B for complete set of stimuli). Samples for the flower items are included in Table 2.1. Ratings were collected on a 7-point scale with anchors at 1 ("Very Poor Explanation") and 7 ("Excellent Explanation").

Before concluding the experiment, participants completed an attention and memory check question that served as the basis for participant exclusion.[7] Finally, participants were asked to report their age and sex.

Results

Overall rule reporting. Participants reported finding an average of 1.23 patterns *(SD = 1.23, min = 0, max = 9)* that they reported accounted for an average of 8.18 exemplars *(SD = 3.13, min = 0, max = 12)*. Reported patterns were coded for mention of the better rule and/or the worse rule.

Better rule reporting. To test whether explanation prompts affected discovery of the better rule (100 percent or 83 percent, depending on pattern-type), and whether effects differed across pattern-type (see Figure 2.2), we conducted a logistic regression predicting whether participants *discovered the better rule*

Table 2.1 Average rule ratings by condition

Rule Type	Flower Rule Text	Explainers-Ideal Rule Condition	Thinkers-Ideal Rule Condition	Explainers-No Ideal Rule Condition	Thinkers-No Ideal Rule Condition
Better	Because SOMP flowers have two circles in their centers, and THONT flowers have one circle in their centers.	6.07 (1.72)	5.95 (1.84)	4.41 (1.97)	4.32 (2.00)
Worse	Because SOMP flowers have petals which are all the same color, and THONT flowers have petals with many colors.	3.72 (2.19)	3.65 (2.05)	3.05 (1.80)	2.89 (1.73)
Worse + exception	Because SOMP flowers have petals which are all the same color, and THONT flowers have petals with many colors, though there are some exceptions.	4.74 (2.21)	5.03 (2.13)	4.26 (1.97)	4.26 (2.12)
Bad	Because SOMP flowers have black polka-dots on their petals, and THONT flowers have no black polka-dots on their petals.	1.49 (1.25)	1.43 (1.22)	1.70 (1.36)	1.54 (1.27)

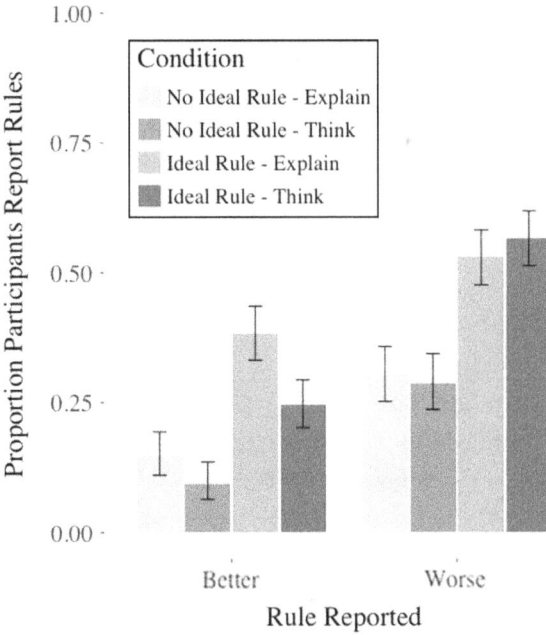

Figure 2.2 The proportion of participants reporting each rule in Experiment 1, as a function of rule type, condition, and prompt. Error bars correspond to 95 percent confidence intervals.

(yes vs. no) by *prompt-type* (explain vs. write thoughts) × *pattern-type* (*ideal rule* vs. *no ideal rule*) × *stimulus-type* (flowers vs. containers vs. objects vs. simple robots vs. complex robots). This revealed a significant effect of prompt-type on reporting the better rule ($\chi^2 = 17.23$, $p < 0.01$), with higher discovery rates for participants prompted to explain. There was also a significant main effect of pattern-type, with more participants reporting the better rule when it accounted for more items ($\chi^2 = 72.38$, $p < 0.01$). The interaction term between prompt-type and pattern-type was not significant ($\chi^2 = 0.63$, $p = 0.43$). The interaction term between prompt-type and stimulus-type was also not significant ($\chi^2 = 6.99$, $p = 0.14$).[8] These findings suggest that explaining did indeed facilitate discovery of the better rule, regardless of whether the better rule was ideal, and across a range of different stimulus types.

The results of this analysis are consistent with the hypothesis that when explaining, people seek simple and exceptionless rules, but that in the course of doing so, they are likely to discover "good" rules that may nonetheless fall short of this ideal. To further investigate this pattern of results, we ran additional logistic regressions for the *ideal rule* condition and *no ideal rule* condition separately.

We found that explainers reported the better rule significantly more often than those who wrote their thoughts within the *ideal rule* pattern-type condition ($\chi^2 = 15.53$, $p < 0.01$) and also within the *no ideal rule* condition ($\chi^2 = 3.94$, $p = 0.05$).[9] These results further support the claim that engaging in explanation can assist people in discovering the best available rule, even when it is imperfect.

Worse rule reporting. Previous studies have found that prompting participants to explain can sometimes decrease worse rule reporting relative to a control condition (e.g., Edwards, Williams, and Lombrozo, 2013; Williams and Lombrozo, 2010, 2013). To analyze worse rule reporting, we ran another logistic regression: *discovered the worse rule* (yes vs. no) by *prompt-type* (explain vs. write thoughts) × *pattern-type* (*ideal rule* vs. *no ideal rule*) × *stimulus-type* (flowers vs. containers vs. objects vs. simple robots vs. complex robots). The effect of prompt-type was not significant ($\chi^2 = 0.39$, $p = 0.53$). The effect of pattern-type was significant ($\chi^2 = 84.79$, $p < 0.01$): participants reported the 83 percent worse rule more often than the 66 percent worse rule. However, the interaction between prompt-type and pattern-type was not significant ($\chi^2 = 1.00$, $p = 0.32$).[10] These findings suggest that while explaining improved discovery of the better rule, it did so at no cost to discovery of the worse rule.

Rule Ratings. To analyze rule ratings (see Table 2.1), we first confirmed that participants were attentively engaged in the task by verifying that ratings for the bad rule were significantly lower than those for the other three options. Using *t*-tests comparing each of the three "good" options against the bad rule within each of the four conditions revealed a significant difference in each case, even using a Bonferroni correction for multiple comparisons.

To analyze ratings for the three good rules, we performed an ANOVA with prompt-type (2: explain, write thoughts) and pattern-type (2: *ideal rule*, *no ideal rule*) as between-subjects factors, and rule rated (3: better rule, worse rule, worse rule acknowledging exceptions) as a within-subjects factor. This analysis revealed no main effect of prompt-type, $F(1, 992) < 0.01$, $p = 0.97$, a significant main effect of pattern-type, $F(1, 992) = 133.21$, $p < 0.01$, and a significant effect of rule rated, $F(2, 1984) = 292.11$, $p < 0.01$. The main effects of pattern-type and rule rated were qualified by a significant interaction, $F(2, 1984) = 24.75$, $p < 0.01$.[11] Not surprisingly, the better rule was rated more highly when it accounted for 100 percent of cases than when it accounted for 83 percent of cases, $t(923) = -13.61$, $p < 0.01$, consistent with our assumption that explanatory evaluation favors patterns without exceptions. We also found that the worse rule was rated more highly when it accounted for more cases, whether the rule did, $t(982) = -4.68$, $p < 0.01$, or did not, $t(994) = -5.80$, $p < 0.01$, mention exceptions.

This is consistent with our finding that explaining also favors the discovery of patterns that account for more cases, even when both fall short of 100 percent. However, the gap in ratings between the "better" 83 percent and 100 percent rules (1.64 points on a 7-point scale) was greater than that between the "worse" 66 percent and 83 percent rules, both when the rule did (0.63 points) or did not (0.71 points) mention exceptions, accounting for the significant interaction and also suggesting that there may be something special about a rule without any exceptions.

Discussion

The results of Experiment 1 both replicate and extend prior research. Consistent with prior research, we found that a prompt to explain facilitated discovery of a subtle, exceptionless rule. Going beyond prior research, we also found that a prompt to explain facilitated the discovery of a subtle rule that involved exceptions, albeit *fewer* exceptions than a more salient alternative. This helps resolve the puzzle with which we began. On the one hand, seeking explanations seems to push learners to achieve an explanatory ideal, which involves simple, exceptionless generalizations. (Indeed, our rule rating results suggest that such generalizations are highly valued.) But on the other hand, real-world domains rarely support the realization of this ideal. We find that even in a domain where the ideal cannot be attained, engaging in explanation may be useful because it pushes learners to go beyond the obvious in search of a "better"—albeit imperfect—regularity.

Experiment 2: Extension to an easier learning environment

In Experiment 2, we sought to replicate and extend the results of Experiment 1. Specifically, the experiment considers whether the effects observed in Experiment 1 will generalize to a context in which the task of discovery is simplified by making the defects of the "worse" rule, and the features that support the "better" rule, easier to identify. One possibility is that even in an easier learning task, the effects of explanation will continue to surpass those of our control condition. But another possibility is that by making it easier for participants in the control condition to "go beyond the obvious," we will boost their performance to a level comparable to that of participants prompted to explain.

How might the subtler pattern be made "more obvious" in a relevant way? Experiment 2 altered the difficulty of the learning task by using a different presentation format. In Experiment 2, items were presented in a more structured array, and participants were asked to consider all items in a category together rather than studying each item independently. Previous work suggests that presentation format can have a significant effect on category learning (Meagher et al., 2017), and that increasing the ease with which (or the rate at which) category members are compared can also affect learning (Edwards et al., 2019; Lassaline and Murphy, 1998). There's also evidence that explanation may help, in part, by drawing attention to category-wide statistical properties, and by encouraging participants to focus on the contrast between one category and the other (Edwards et al., 2019; Chin-Parker and Bradner, 2017). One might therefore expect the presentation format in Experiment 2 to mimic some of the benefits of explanation by making category-wide properties and diagnostic features more salient, thereby rendering the inferior rule more obviously inferior, and the more subtle pattern less difficult to detect.

In sum, Experiment 2 contrasts two hypotheses: that effects of explanation (relative to a control condition) are fairly stable across variation in the difficulty of a learning task, and the alternative that the effects of explanation (relative to a control condition) are moderated by task difficulty. In an easier learning task, even participants in a control condition may readily go beyond the obvious, rendering a prompt to explain somewhat superfluous. Such a moderating effect could shed light on whether and why effects of explanation could vary as a function of the domain, data, and tools for analysis available to a scientist or everyday learner.

Method

Participants

The sample for Experiment 2 consisted of 1470 adults recruited as in Experiment 1.[12]

Materials

Stimuli were the same as those used in Experiment 1.

Procedure

The experiment consisted of a study phase and a rule reporting phase that were very similar to those of Experiment 1. However, there were two key differences. Rather than randomly arranging the twelve items into a 3 × 4 grid, items were grouped by category in two 3 × 2 groups. Additionally, rather than being asked about each item individually in a sequential manner, participants were given three minutes to study each of the two categories, with a prompt to either explain or write their thoughts about all of the items in a category together.

Results

Better rule reporting. To test whether explainers reported the better rule more frequently than thinkers (see Figure 2.3), we ran a logistic regression predicting whether participants *discovered the better rule* (yes vs. no) by *prompt-type* (explain vs. write thoughts) × *pattern-type* (ideal rule vs. no ideal rule) × *stimulus-type* (flowers vs. containers vs. objects vs. simple robots vs. complex robots). We found no effect of prompt-type ($\chi^2 = 0.30$, $p = 0.58$), suggesting that in this context, explaining does not facilitate the discovery of the better rule. There was also a significant effect of pattern-type ($\chi^2 = 73.79$, $p < 0.01$), with higher discovery rates when the better rule accounted for more cases.[13]

To further investigate why there was a significant effect of prompt-type on better rule discovery in Experiment 1 but not in Experiment 2, we ran a logistic regression predicting whether participants *discovered the better rule* (yes vs. no)

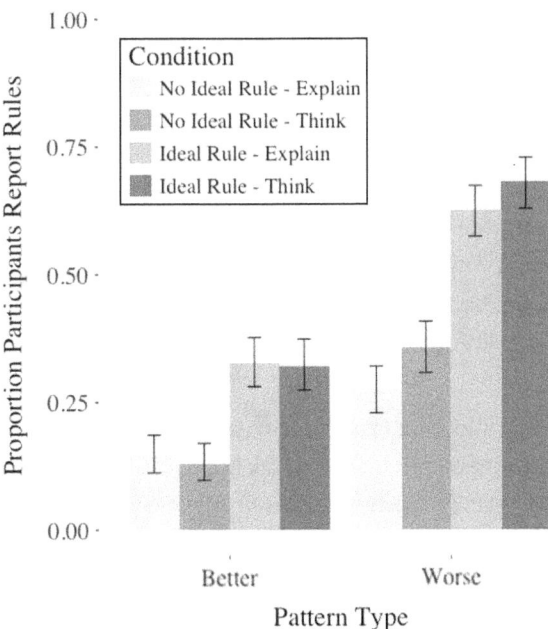

Figure 2.3 The proportion of participants reporting each rule in Experiment 2, as a function of rule type, condition, and prompt. Error bars correspond to 95 percent confidence intervals.

by *experiment number* (1 vs. 2) × *prompt-type* (explain vs. write thoughts) × *pattern-type* (ideal rule vs. no ideal rule) × *stimulus-type* (flowers vs. containers vs. objects vs. simple robots vs. complex robots). The goal of doing so was to identify whether changing the presentation format (as we did in Experiment 2) had a differential effect on explainers versus participants in the control group. We found a significant interaction between experiment number and prompt-type ($\chi^2 = 10.83$, $p < 0.01$). This interaction suggests that the change in presentation format did in fact impact explainers and control participants differently. To explore the nature of this difference, we ran additional logistic regressions for explainers and control participants separately. The change in presentation format between Experiments 1 and 2 only had a significant effect on better rule discovery for control participants ($\chi^2 = 7.09$, $p = 0.01$) and not for explainers ($\chi^2 = 1.76$, $p = 0.18$). It therefore appears that the changes to the presentation format in Experiment 2 (simultaneous presentation in separated categories) allowed control participants to perform more like explainers, perhaps by making the need to "go beyond the obvious" less effortful.

Worse rule reporting. To test whether prompt-type influenced the discovery of the worse rule, we ran another logistic regression predicting whether participants *discovered the worse rule* (yes vs. no) by *prompt-type* (explain vs. write thoughts) × *pattern-type* (ideal rule vs. no ideal rule) × *stimulus-type* (flowers vs. containers vs. objects vs. simple robots vs. complex robots). There was a significant effect of prompt-type ($\chi^2 = 6.60$, $p = 0.01$), suggesting that explaining *inhibited* discovery of the worse rules in this task (see Figure 2.3). There was also a significant effect of pattern-type ($\chi^2 = 173.59$, $p < 0.01$), with more frequent discovery of the worse pattern when it accounted for more cases.[14] Unlike Experiment 1, this suggests that explaining can result in lower detection or reporting of regularities that are superseded by better alternatives.

Discussion

Comparing the results of Experiment 1 to those of Experiment 2 suggests that when a learning environment makes it easier to recognize the flaws of a suboptimal pattern and to identify the features that support a better alternative, control participants (i.e., those who are *not* prompted to explain) receive a disproportionate benefit. By contrast, participants who are prompted to explain more often succeed in going beyond the obvious to find a better but more subtle pattern *whether or not* a learning environment makes it easy to do so. These findings reinforce a lesson from prior research: that the magnitude of the benefits of explanation (relative to control conditions) can be moderated by a variety of factors (Kon and Lombrozo, in prep; Williams, Lombrozo, and Rehder, 2013). But they also go beyond prior research in finding that task difficulty may be one of these moderating factors. Generalizing to science, we might expect the drive for explanation to have an especially pronounced effect on scientific discovery when the dominant ways of representing the relevant data or phenomena do not already support the alignment of relevant features or comparisons across relevant distinctions.

General discussion

Across two studies, we find support for a potential resolution to the puzzle with which we began. On the one hand, scientists are often driven to achieve an explanatory ideal with a prominent role for exceptionless generalizations and theories that support simple explanations. On the other hand, regularities in the natural world quite often have exceptions, and simple explanations are not always forthcoming. Our findings suggest that the process of seeking ideal explanations may be beneficial because it supports discovery, and that these beneficial effects on discovery are not restricted to the ideal case; explaining can facilitate the discovery of subtle patterns even when those patterns do not account for all cases. This finding is broadly consistent with the idea of "Explaining for the Best Inference" (EBI) introduced by Wilkenfeld and Lombrozo (2015); the *process* of seeking explanations can sometimes be beneficial because it has positive downstream consequences on what we learn and infer.

Needless to say, our artificial learning tasks are a poor match to real scientific practice, and our classification rules are a poor match to rich scientific explanations. The research we review and present here is no substitute for naturalistic studies of real scientific advance. That said, we expect the learning mechanisms documented here to apply quite broadly. For example, findings concerning the effects of explanation in artificial classification tasks (Williams and Lombrozo, 2010) have been replicated with property-generalization tasks that involve meaningful causal explanations (Kon and Lombrozo, in prep). The core phenomena found with adults have also been successfully replicated with preschool-aged children (Walker et al., 2014; Walker, Bonawitz and Lombrozo, 2017). These findings suggest that effects of engaging in explanation are fairly widespread and baked into our explanatory activities from a young age. While science undoubtedly involves a refinement of these widespread explanatory tendencies, we expect a great deal of continuity to be maintained nonetheless.

Beyond the lack of scientific realism, other limitations of these studies should be acknowledged. Our participant pool was restricted to online participants within the United States, our learning tasks occurred over a short time scale, and participants were almost certainly more motivated to receive their pay than to uncover the structure of our artificial worlds. Moving forward, it will be important to pursue research that preserves the experimental control of the studies we present here while simultaneously overcoming these limitations.

Zooming out, our findings support a functionalist approach to scientific explanation (Lombrozo, 2011). On this view, explanation is crucial to science because it serves an instrumental role. By pursuing explanations of the natural world, we're more likely to generate discoveries and develop theories that in turn improve our strategic position in the world, satisfying Hempel's first motivation for science by pursuing the second.

Appendix A: Better and worse rules for all stimulus types

- Flowers
    - 100 percent ideal & 83 percent no ideal—SOMP flowers have two concentric circles in their centers; THONT flowers have one circle in their center.
    - 83 percent ideal & 66 percent no ideal—The petals of SOMP flowers are mostly one color; the petals of THONT flowers are mostly rainbow-colored.

Figure 2.4 Flowers.

- Containers
  - 100 percent ideal & 83 percent no ideal—ANDRAK containers rest on platforms that are larger than their openings; ORDEEP containers rest on platforms that are smaller than their openings.
  - 83 percent ideal & 66 percent no ideal—ANDRAK containers were mostly tall and narrow; ORDEEP containers were mostly short and wide.

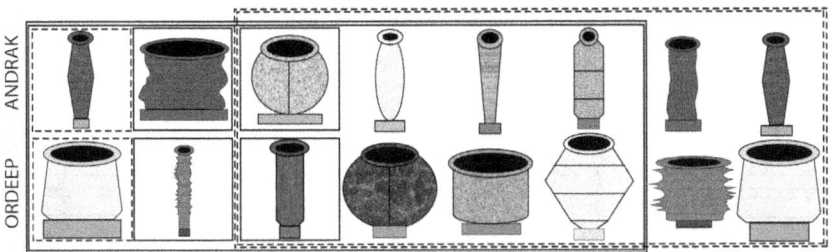

Figure 2.5 Containers.

- Objects
  - 100 percent ideal & 83 percent no ideal—TRING objects have their larger portion on the top; KRAND objects have their larger portion on the bottom.
  - 83 percent ideal & 66 percent no ideal—TRING objects mostly have vertical lines dividing their sections with and without dots; KRAND objects mostly have diagonal lines dividing their sections with and without dots.

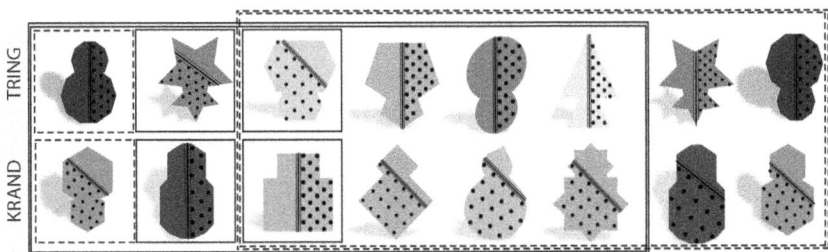

Figure 2.6 Objects.

- Simple robots
  - 100 percent ideal & 83 percent no ideal—the bottom of the DRENT robots' feet were flat; the bottom of the GLORP robots' feet were pointed.
  - 83 percent ideal & 66 percent no ideal—the bodies of most DRENT robots were round; the bodies of most GLORP robots were square.

Figure 2.7 Simple robots.

- Complex robots
  - 100 percent ideal & 83 percent no ideal—the bottom of the DRENT robots' feet were flat; the bottom of the GLORP robots' feet were pointed.
  - 83 percent ideal & 66 percent no ideal—
    1) The bodies of most DRENT robots are round, and the bodies of most GLORP robots are square.
    2) The antennae of most DRENT robots are curled, and the antennae of most GLORP robots are straight.
    3) Most DRENT robots have stars on their hands, and most GLORP robots have nothing at the ends of their arms.
    4) Most DRENT robots have no dot on their chest, and most GLORP robots have a chest dot.
    5) Most DRENT robots have their colored sections split evenly down the middle, and most GLORP robots have a checkered pattern.

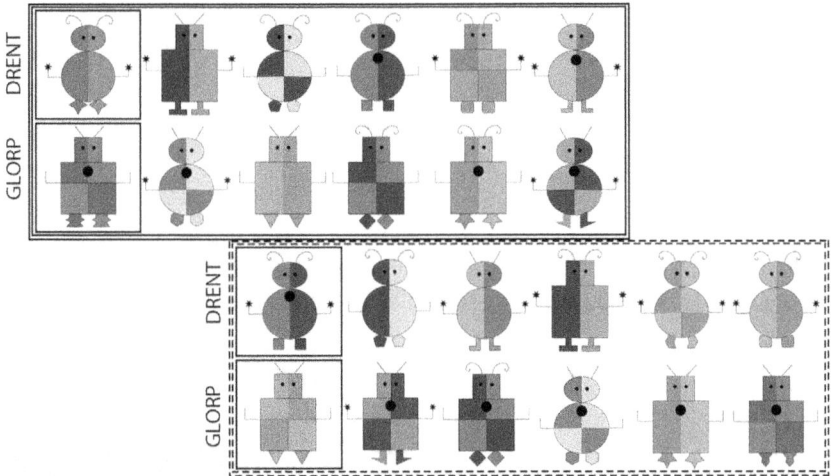

Figure 2.8 Complex robots.

Appendix B: Rated rules

- Better rule:
  - Because SOMP flowers have two circles in their centers, and THONT flowers have one circle in their centers.
  - Because ANDRAK containers rest on platforms that are larger than their openings, and ORDEEP containers rest on platforms that are smaller than their openings.
  - Because TRING objects have a larger shape on top of a smaller shape, and KRAND objects have a smaller shape on top of a larger shape.
  - Because GLORP robots have feet that are pointy along the bottom, and DRENT robots have feet that are flat along the bottom.
- Worse rule:
  - Because SOMP flowers have petals which are all the same color, and THONT flowers have petals with many colors.
  - Because ANDRAK containers have thin bodies, and ORDEEP containers have wide bodies.

- Because TRING objects have a vertical line dividing their sections with and without dots, and KRAND objects have a diagonal line dividing their sections with and without dots.
- Because GLORP robots have square-shaped bodies, and DRENT robots have circular bodies.

- Worse rule + exception:
    - Because SOMP flowers have petals which are all the same color, and THONT flowers have petals with many colors, though there are some exceptions.
    - Because ANDRAK containers have thin bodies, and ORDEEP containers have wide bodies, though there are some exceptions.
    - Because TRING objects have a vertical line dividing their sections with and without dots, and KRAND objects have a diagonal line dividing their sections with and without dots.
    - Because GLORP robots have square-shaped bodies, and DRENT robots have circular bodies, though there are some exceptions.

Notes

1 Some historical analyses do aim to chart psychological aspects of scientific discovery (e.g., Gentner et al., 1997), use observational/ethnographic methods with qualitative and quantitative analyses to better understand scientific practice (e.g., Dunbar, 1997), or consider how science works from a cognitive scientific perspective (e.g., Proctor and Capaldi, 2012; Thagard, 2012). To our knowledge, however, this research has not focused on how psychological features of our drive for explanations affect scientific advance.

2 It's worth clarifying a feature of our nomenclature regarding classification rules. In labeling a rule as 66 percent, 83 percent, or 100 percent, we are highlighting the percentage of training items captured by an unqualified generalization (e.g., "Thont flowers have a single circle in their centers"); we do not intend the percentage to be built-in to form a probabilistic classification rule (e.g., "Thont flowers have a single circle in their centers with 66 percent probability"), in which case the "exception" items would arguably still fall under the generalization.

3 The mean age of participants was 34 ($SD = 11$, min = 18, max = 74); 510 participants identified as male and 857 as female. Initially, we collected a sample of 1309 participants (before exclusions); however, when we analyzed these responses, the results were inconclusive, as we explain in footnote 9. We therefore collected data from additional participants. Analyses correspond to the full sample, but the patterns were the same within each subsample.

4 An additional 1007 participants failed attention or memory checks (see footnote 6) and were therefore excluded from analyses. We indicate any cases in which these exclusions affect the statistical significance of results.

5 Data on the complex robot stimuli were collected separately from the other four stimulus-types, and the stimuli and procedure varied slightly. Specifically, the complex robot stimuli contained five equally good "worse" rules (rather than only one) in addition to one better rule, and participants did not complete the rule rating phase. We combine the data here because the experimental questions and results were the same.

6 Specifically, participants were told "we're interested in any patterns that you noticed that might help differentiate SOMPS and THONTS. For example, did most or all of the SOMPS you studied tend to have one property, and most or all of the THONTS you studied have another property? We're going to ask you to list all of the patterns (differences between SOMPS and THONTS) that you noticed, one at a time. PLEASE REPORT ANY PATTERNS THAT YOU NOTICED, EVEN IF THEY WEREN'T PERFECT AND EVEN IF YOU DON'T THINK THEY'RE IMPORTANT." This language, adapted from Edwards, Williams, and Lombrozo (2013), was employed to encourage participants to report the worse rule (83 percent in the *ideal rule* condition, and 66 percent in the *no ideal rule* condition) even if they thought it was incidental or superseded by the better rule (100 percent in the *ideal rule* condition, and 83 percent in the *no ideal rule* condition).

7 This consisted of a fairly long passage that asked them to select "None of these objects look familiar" and to write in the category of the item they recognized. Specifically, it said "look at the following images and select the one that you have studied in previous questions. In the text box next to that image, please also type in whether you think that it is a [category 1] or a [category 2]. It is important for us to know whether our participants are paying attention and are reading all of the instructions, so if you are reading this, what we actually want you to do is to select 'None of these objects look familiar,' and in the corresponding text box to write in whether the image you recognize from the other options is a [category 1] or a [category 2]." By selecting the instructed button, participants indicated they had been reading instructions, and by correctly reporting the category of the item they recognized, participants indicated that they attended to the stimuli in the primary task.

8 There was also a significant main effect of stimulus-type ($\chi^2 = 112.98$, $p < 0.01$), and a significant interaction between pattern-type and stimulus-type ($\chi^2 = 13.72$, $p < 0.01$).

9 We initially collected a smaller sample size, but the statistical analyses were inconclusive. Specifically, we found the expected effect of explanation (with more participants reporting the better rule when prompted to explain), but we also (a) failed to find an interaction between pattern-type and prompt-type,

suggesting that the effects of explanation were comparable across the *ideal* and *no ideal* rule conditions, and (b) failed to find a significant effect of the explanation prompt when restricting analysis to the *no ideal rule* condition, suggesting that explanation did *not* have an effect under these conditions. Because (a) and (b) supported different conclusions, we decided to collect additional data. It is worth noting that while increasing the sample size did change the statistical significance of the effect of explanation within the *no ideal rule* condition, the proportions of participants reporting the rules remained fairly unchanged by the increased sample size (approximately 15 percent of the explainers reported the imperfect better rule in both the initial and increased sample, approximately 10 percent of control participants reported the imperfect better rule in the initial sample, and approximately 9 percent reported it in the increased sample). This suggests that the initial sample was simply underpowered.

10 The effect of stimulus-type was also significant ($\chi^2 = 57.08$, $p < 0.01$); no interactions were significant (without exclusion criteria, the interaction between pattern-type and stimulus-type was significant ($\chi^2 = 13.79$, $p = 0.01$)).

11 Without exclusion criteria, there is also a significant interaction between prompt-type and rule rated ($\chi^2 = 4.00$, $p = 0.02$)

12 An additional 1048 participants failed attention or memory checks and were therefore excluded from analyses. The statistical significance of results are unchanged unless noted when these participants are included. The mean age of participants was 35 ($SD = 11$, min = 18, max = 79); 1020 participants identified as male and 1487 as female. As in Experiment 1, initially a smaller sample was collected (949), but this was increased to keep approximately the same number of participants across the two experiments.

13 There was also a significant effect of stimulus-type ($\chi^2 = 104.18$, $p < 0.01$). No interaction was significant (without exclusion criteria, there was a significant interaction between pattern-type and stimulus-type ($\chi^2 = 10.24$, $p = 0.04$)).

14 There was also a significant effect of stimulus-type ($\chi^2 = 59.44$, $p < 0.01$). No interaction was significant (without exclusion criteria, there was also a significant interaction between prompt-type and stimulus-type ($\chi^2 = 13.60$, $p = 0.01$)).

References

Chi, M. T. H. (2000). Self-explaining expository texts: The dual processes of generating inferences and repairing mental models. *Advances in Instructional Psychology*, 5: 161–238.

Chi, M. T. H., Bassok, M., Lewis, M., Reimann, P., and Glaser, R. (1989). Self-explanations: How students study and use examples in learning to solve problems. *Cognitive Science*, 13: 145–82.

Chin-Parker, S. and Bradner, A. (2017). A contrastive account of explanation generation. *Psychonomic Bulletin and Review*, 24(5): 1387–97.

Dunbar, K. (1997). How scientists think: Online creativity and conceptual change in science. In T. B. Ward, S. M. Smith and S. Vaid (Eds.), *Conceptual Structures and Processes: Emergence Discovery and Change* (pp. 461–93). Washington: American Psychological Association.

Edwards, B. J., Williams, J. J., and Lombrozo, T. (2013). Effects of explanation and comparison on category learning. In M. Knauff, M. Pauen, N. Sebanz, and I. Wachsmuth (Eds.), *Proceedings of the 35th Annual Conference of the Cognitive Science Society* (pp. 406–11). Austin: Cognitive Science Society.

Edwards, B. J., Williams, J. J., Gentner, D., and Lombrozo, T. (2019). Explanation recruits comparison: Insights from a category-learning task. *Cognition*, 185: 21–38.

Fonseca, B. and Chi, M. T. H. (2011). Instruction based on self-explanation. In R. Mayer and R. Alexander (Eds.), *Handbook of Research on Learning and Instruction*. New York: Routledge.

Gentner, D., Brem, S., Ferguson, R. W., Markman, A. B., Levidow, B. B., Wolff, P., and Forbus, K. D. (1997). Analogical reasoning and conceptual change: A case study of Johannes Kepler. *The Journal of the Learning Sciences*, 6(1): 3–40.

Harré, R. (1985). *The Philosophies of Science*. Oxford: Oxford University Press.

Hempel, C. (1962). Explanation in science and history. In R. G. Colodny (Ed.), *Frontiers of Science and Philosophy* (pp. 7–33). London: Allen & Unwin.

Kon, E. and Lombrozo, T. (in prep). Why explainers take exception to exceptions.

Lassaline, M. E. and Murphy, G. L. (1998). Alignment and category learning. *Journal of Experimental Psychology: Learning, Memory, and Cognition*, 24(1): 144–60.

Legare, C. H. and Lombrozo, T. (2014). Selective effects of explanation on learning during early childhood. *Journal of Experimental Child Psychology*, 126: 198–212.

Lombrozo, T. (2011). The instrumental value of explanations. *Philosophy Compass*, 6: 539–51.

Lombrozo, T. (2012). Explanation and abductive inference. In K. J. Holyoak and R. G. Morrison (Eds.), *The Oxford Handbook of Thinking and Reasoning*. Oxford: Oxford University Press.

Lombrozo, T. (2016). Explanatory preferences shape learning and inference. *Trends in Cognitive Science*, 20(10): 748–59.

Meagher, B. J., Carvalho, P. F., Goldstone, R. L., and Nosofsky, R. M. (2017). Organized simultaneous displays facilitate learning of complex natural science categories. *Psychonomic Bulletin and Review*, 24(6): 1987–94.

Pitt, J. C. (1988). *Theories of Explanation*. Oxford: Oxford University Press.

Proctor, R. W. and Capaldi, E. J. (Eds.) (2012). *Psychology of Science: Implicit and Explicit Processes*. Oxford: Oxford University Press.

Rozenblit, L. and Keil, F. (2002). The misunderstood limits of folk science: An illusion of explanatory depth. *Cognitive Science*, 26(5): 521–62.

Thagard, P. (2012). *The Cognitive Science of Science: Explanation, Discovery, and Conceptual Change*. Cambridge: MIT Press.

Walker, C. M., Bonawitz, E., and Lombrozo, T. (2017). Effects of explaining on children's preference for simpler hypotheses. *Psychonomic Bulletin and Review*, 24(5): 1538–47.

Walker, C.M., Lombrozo, T., Legare, C., and Gopnik, A. (2014). Explaining prompts children to privilege inductively rich properties. *Cognition*, 133: 343–57.

Wilkenfeld, D. A. and Lombrozo, T. (2015). Inference to the best explanation (IBE) versus explaining for the best inference (EBI). *Science and Education*, 24(9–10): 1059–77.

Williams, J. J. and Lombrozo, T. (2010). The role of explanation in discovery and generalization: Evidence from category learning. *Cognitive Science*, 34(5): 776–806.

Williams, J. J. and Lombrozo, T. (2013). Explanation and prior knowledge interact to guide learning. *Cognitive Psychology*, 66: 55–84.

Williams, J. J., Lombrozo, T., and Rehder, B. (2013). The hazards of explanation: Overgeneralization in the face of exceptions. *Journal of Experimental Psychology: General*, 142(4): 1006–14.

3

The Challenges and Benefits of Mechanistic Explanation in Folk Scientific Understanding

Frank Keil

Experimental philosophy has emerged as an important new interdisciplinary approach in cognitive science. One of most interesting features of this new development is that very diverse areas of philosophy can contribute to experimental studies. Thus, many experimental studies build on moral philosophy, the philosophy of mind, and the philosophy of language. Somewhat less common are studies building on the philosophy of science. But this is starting to change. So, for example, in recent years, a number of experimental investigations draw on philosophical research in order to understand which features make for cognitively appealing explanations (see, for example, Keil, 2006; Lombrozo, 2007; Hussak and Cimpian, 2015; Johnston et al., 2018; and Kon and Lombrozo, this volume).

One central question in this new area concerns how humans with limited cognitive capacities are able to make progress in understanding the causal structure of a world of nearly unbounded complexity. Scientists must make various forms of simplifications, idealizations, and abstract models to reason effectively about natural phenomena. They also frequently engage in outsourcing part of their cognitive loads by relying on those with much greater expertise in other areas. These abilities to simplify problems and outsource the cognitive labor seem to come naturally to many scientists, so naturally, in fact, that they may be abilities shared with laypeople and even children. Studying such nonexperts may in turn feed back to the philosophy of science in order to help us better understand how scientists make explanatory progress.

In our research, we frequently turn to young children to see if they have relevant foundational skills that may be elaborated on throughout life while also perhaps creating certain systematic distortions in cognition and understanding that can persist into adulthood and even in the minds of professional researchers. Recently, we have focused on one facet of this problem that has become a major new area of scholarship in the philosophy of science, namely, mechanistic explanation (see, for example, Craver and Kaplan, 2018; Felline, 2018; Glennan, 2017; Glennan and Illari, 2018; Kästner and Andersen, 2018; Roe and Baumgaertner, 2017; Rosenberg, 2018). Philosophers have started to characterize the distinctive nature of mechanistic explanations and understanding. Such characterizations make it possible for empirical researchers to ask whether mechanistic forms of understanding and explanation have a distinctive cognitive role in laypeople and in the origins of understanding in children.

This chapter begins by briefly characterizing how mechanistic explanations have emerged as a major topic of study in the philosophy of science. It turns to two phenomena that seem to create a tension: the apparent failure at all ages to remember mechanisms, and the extraordinary amount of interest in mechanisms shown by young children. A possible resolution of this tension is suggested by way of recognizing the importance of higher-level abstractions that arise from exposure to information about mechanism. Experimental support for this suggestion is then provided. The powerful utility of such abstractions in supporting cognitive outsourcing is then discussed along with some attendant illusions. Finally, implications for future experimental and conceptual work and their interactions are considered.

The rise of mechanism in the philosophy of science

The recent consensus on the importance of mechanistic understanding in the sciences has arisen primarily from considerations in biology and cognitive science (e.g., Bechtel, 2011; Bechtel and Richardson, 1993; Machamer, Darden, and Craver, 2000; Thagard, 2000). This emphasis on biology is interesting because it might seem that mechanical systems are more intrinsic to physics, where mechanical causation is seemingly more iconic. However, biological systems tend to have more discernible hierarchical causal structures in which subcomponents with apparent functional roles causally interact in a manner that helps to explain higher-order properties of the system. In addition, the very use of causal language in foundational physics has been controversial (Frisch, 2014; Russell, 1912). In view of such considerations, biology has been the region of science in which philosophers have most systematically explored the nature and role of mechanistic explanations, a development that may be related to an increasing acceptance of teleological language in biology (e.g., Canfield, 1964; Wright, 1976). Mechanistic explanations are different from teleological language (e.g., birds fly because their breast muscles enable them to flap their wings and provide thrust that moves them through the air while also providing an airfoil shape that provides lift vs. birds can fly because flying allowed them to avoid prey and seek out new food sources), but as seen in the prior example, functional language (a close relative of teleological language) is often present in mechanistic explanations. Indeed, many accounts in biology are more akin to descriptions of how complex artifacts work than they are to descriptions of nonliving natural systems. This focus on mechanism has been common practice in science for centuries (Dolnick, 2011; Faber, 1986).

Although mechanistic explanation has been characterized in a variety of distinct ways, according to the most influential proposals—such as those of Bechtel, Machamer, Darden, and Craver—the following properties are attributed to them:

First, there is the notion of a layered structure in which operations at one level are unpacked into causal interactions at a lower level. More complex cases also exist that involve cycles and networks (e.g., Bechtel, 2017a; in press). Here, however, we restrict ourselves to more straightforward hierarchies. In addition, it is assumed that in most cases, causal processes segregate so that there are distinct levels. This assumption connects with a psychological bias to reason about events at the same level of causal analysis and not shift up and down repeatedly across levels (Johnson and Keil, 2014).

Second, most often the components of the lower level are broken down into subassemblies performing specific functions that interact in coherent ways with other subassemblies to help explain higher-order functions. Thus, most mechanistic accounts in the biological sciences discuss the roles that each constitutive component plays. It is much less common to refer to lower-level components as part of a causal chain but having no distinctive functions.

Third, these patterns of interactions are assumed to provide insight into how a higher-level process becomes manifested. That is, a full specification of how components work together at one level enables one to make more refined inductions at the next level up. Such insights can occur even when the next level down remains opaque.

Finally, there seems to be a persistent sense of work, the transfer of energy from one place to another, that is performed at one level in service of a higher level (Bechtel, 2017b). This has been described as constraining or redirecting the flow of free energy in a system. Thus, mechanisms typically involve notions of energy transmission and direction.

A concrete example is the case of mitosis and how the entire process is broken down into subprocesses with distinctive roles such as prophase, metaphase, anaphase, and telophase. Each of these phases in turn unpacks into subprocesses with their own distinctive roles. For example, during the metaphase, a crucial series of steps governs the alignment of chromosomes through a complex process that requires extraordinary precision to avoid genetic anomalies (Ali et al., 2017). Pragmatically, this often means that different specialists are required to spend their entire careers focusing on just one of these subprocesses. These hyperspecializations in turn mean that there must be ways in which distinct specialists come to fruitfully rely on each other.

For highly complex artifacts, such as computer systems or space vehicles, the same hierarchical structures and divisions of cognitive labor are also commonplace. These cases are in sharp contrast to the simple tools and artifacts that were the only artifacts for much of human history. This contrast between simple and complex artifacts and the roles of mechanism in both has led to a recent body of research exploring how young children come to differentiate these two kinds of artifacts (Ahl, 2018).

From a cognitive science perspective, one of the most fascinating questions about mechanistic understanding concerns what role it could possibly play in folk science. The challenge is that mostly people seem to have absolutely dismal levels of mechanistic understanding. In fact, those understandings are so limited that they seem to be completely devoid of relevant details. It therefore might be argued that mechanistic understanding plays no role in the cognitive lives of ordinary people. Yet, in the last decade or so, research with children suggests that they might be particularly inclined to seek out mechanistic explanations (Chouinard, Harris, and Maratsos, 2007; Kurkul and Corriveau, 2017; Frazier, Gelman, and Wellman, 2009, 2016). Why would they engage in such behavior if, in the end, they don't seem to remember anything about the mechanism? We have been conducting a series of empirical studies to see if there is an answer to this puzzle.

Apparent mechanistic ignorance

It is important to document the two aspects of the puzzle, namely, the apparent absence of mechanistic understanding in adults (let alone children), and whether there is indeed nonetheless a fascination with mechanisms in young children.

With respect to adult ignorance of mechanisms, several studies indicate major gaps in understanding. One especially notable example concerns how adults understand the workings of simple bicycles (Lawson, 2006). When shown schematic drawings of alleged bicycles, adults frequently choose as working bicycles drawings that depict mechanistically impossible cases. For example, many adults identify as a working bicycle a case where the chain connects the rear and front axles in such a manner that would make it impossible to steer the bicycle.

This mechanistic ignorance has been repeatedly shown across a wide range of domains in the sciences and engineering (Fisher and Keil, 2016). It may be somewhat surprising to learn about such high levels of ignorance because of a second phenomenon related to illusions of understanding. Across a wide variety of contexts and content domains, people display an illusion of explanatory depth (IOED), in which they think they understand the workings of the world in far more detail and depth than they really do (Alter, Oppenheimer, and Zemla, 2010; Rozenblit and Keil, 2002). Whether it is a flush toilet or human heart, people erroneously assume that they have a credible working model in their heads, whereas when queried, they turn out not to have one at all. Children are vastly worse off on both accounts: they have even more empty understandings and substantially larger illusions of explanatory depth (Mills and Keil, 2004).

Part of the problem seems to be a form of decay neglect in which adults and children alike have a very poor grasp of the extent to which a mechanistic understanding, even when initially quite strong, rapidly decays in memory (Fisher and Keil, in prep). There seems to be an erroneous assumption that if one has a good grasp on mechanistic details at some point in time, one retains most of that grasp over many years. In reality, even relative experts can lose almost all the details unless they are actively relying on them in their daily cognitive lives.

An early love of mechanisms?

Given adults' dismal understandings of most mechanisms, it is all the more surprising that young children seem to actively seek out information about mechanisms. This has been shown informally in a variety of ways. For example, in one early study, four and five-year-olds were shown in a box in which an object was placed behind a door and, when the door was reopened, a different object was present (Lecompte and Gratch, 1972). In addition to being surprised at this outcome, a large proportion of children immediately explored the back of the box to try to discover how the transformation could have occurred. More recently, researchers have presented children with anomalous situations and then examined what sorts of spontaneous questions children ask in light of the anomaly (Frazier, Gelman, and Wellman, 2009, 2016). The striking finding is that preschoolers and young school children do not seem content with answers until they encounter ones that carry causal information. If presented with an anomaly, such as a video in which an actor turns the lights off with her foot instead of her hand, young children are prone to ask "why" and "how" questions about the situation. If they are merely given simple factual answers—such as, a declaration that the person turned the light off with her foot—the children show dissatisfaction and persist in asking why and how until they get some kind of causal explanation. They are similarly unhappy if they receive a circular explanation as opposed to one that conveys real causal information (Kurkul and Corriveau, 2017).

More broadly, transcripts of children's spontaneous speech in everyday conversations suggest a surge of "why" and "how" questions during the late preschool period (Chouinard, Harris, and Maratsos, 2007; Hickling and Wellman, 2001). It therefore seems that causal inquiry is a compelling and natural practice for children even before they start formal schooling. Although these studies might be interpreted as suggesting that children often search for underlying mechanisms, they do not focus specifically on causal mechanisms per se; rather, they merely focus on explanations that contain causal information of some sort. In view of this, we have conducted several studies to try to understand whether children recognize the distinctive value of mechanistic explanations.

One approach focuses on epistemic inferences. We take as our point of departure adult intuitions that someone who has mechanistic understanding of a phenomenon has deeper and more extensive knowledge than someone who merely knows facts about the same phenomenon (Lockhart et al in press). For example, if told that one person prefers a particular fictitious car brand for mechanistic reasons having to do with how the car works, while another person prefers the car for nonmechanistic reasons (e.g., it has leather seats), adults infer that the mechanistic explainer knows more about cars. Across several studies, we have shown that even five-year-olds have the same intuitions as adults. The preference seems to be specifically for mechanistic information as opposed to other kinds of causal information about the entity that is useful but does not explain how it works (e.g., that it has fabulous cup-holders).

What explains this apparent early quest for mechanisms if the mechanistic information seems to evaporate in all people, especially young children? Our proposal is that exposure to mechanistic explanations provides a particularly powerful route to learning higher-order abstract principles about a given phenomenon that do not decay over time. For example, if given a detailed explanation of the inner workings of a refrigerator, a child may forget those details quite quickly and indeed never fully understand them, but at the same time may acquire a sense of the causal complexity underlying a refrigerator's operation. A now extensive literature shows that even young children are sensitive to higher-level patterns as well as more concrete ones (e.g., Baer and Friedman, 2017; Cimpian and Markman, 2009; Cimpian and Petro, 2014; Kushnir and Gelman, 2016).

In other lines of work, we have shown that by age seven or so, most children have strong intuitions about the causal complexity of many everyday devices, even as they know very little about how they actually work (Kominsky, Zamm, and Keil, 2018). Mechanistic exposure therefore provides a way of acquiring enduring intuitions about the causal complexity needed to make an entity work. More broadly, exposure to mechanisms often leads to representations of "mechanism metadata" (Kominsky, Zamm, and Keil, 2018) even as mechanistic details fade. It is interesting that in the philosophy of science, the new interest in mechanisms has been linked to the discovery of complexity (Bechtel and Richardson, 2010). This raises the important question of whether intuitions about complexity that are driven by exposure to mechanisms have a different character to intuitions about complexity that arise via other means, such as visual information regarding the complexity of an object, or information regarding the sheer number of components that a system has. We are actively exploring this issue.

Complexity is only one small part of the patterns that may be abstracted on the basis of exposure to mechanisms. One may learn that a system is cyclical in nature, or that is it a tipping-point phenomenon, or that it involves many branching causal cascades (e.g., Strickland, Silver, and Keil, 2017). If even young children are able to acquire such higher-level causal intuitions, the question then arises as to the usefulness of those intuitions. We are in the midst of documenting a wide variety of potential high-level abstractions as induced by young children after brief exposure to new mechanistic information (e.g., Trouche et al., 2017). These findings will motivate further studies on the ways in which these abstractions influence other areas of cognition. We now turn to some of those possible influences.

The role of higher-level causal abstractions

Enduring causal abstractions, acquired through exposure to information about mechanisms, may serve a number of roles.

First, causal abstractions allow individuals to make judgments about whether they can achieve the requisite understanding on their own. For example, a sense of complexity strongly influences the extent to which children believe that they might need to defer to others if they are adequately to understand how something works (Kominsky, Zamm, and Keil, 2018). To be sure, younger children tend to underestimate the extent to which they will need to defer to others (Lockhart et al., 2016), but above and beyond this general underestimation, there are clear differences in intuitions about difference that are related to complexity.

Second, causal abstractions allow us to learn about related devices and entities by having a sense of what kinds of information are likely to be most relevant and should be attended to when learning about the device. For example, if one learns that homeostasis plays an important role in one biological system, one may be more inclined to look for it in biological systems but not in artifacts. Similarly, as noted earlier, one might come to expect certain kinds of branching causal architectures in understanding social behaviors, but not in mechanical systems.

Third, causal abstractions provide us with a sense of who the most relevant experts are when one recognizes that one cannot master the system on one's own and needs input from others. In other words, they help children and adults navigate the division of cognitive labor that is present in all cultures. We have shown the use of such abstractions in a series of studies examining how young children decide who knows what in the world around them based on information about one piece of knowledge that they do have (Keil et al., 2008). For example, if they hear that a person has a great deal of knowledge about one physical-mechanical phenomena, they tend to extrapolate expertise to other systems involving bounded objects interacting with each other through motion.

Fourth, causal abstractions provide information that enables us to sense whether or not an alleged expert knows what they're talking about. If someone who claims to understand how something works repeatedly stresses higher-order causal patterns that conflict with those acquired through exposure to the mechanism, that person's testimony might be doubted. We are currently exploring this phenomenon in a teaching paradigm, where we teach children about the mechanisms underlying internal combustion engines and then present them with two elected experts, one of whom describes engines in ways that are mechanistically compatible with an engine's causal mechanics (e.g., the critical importance of synchronized timing) and one who describes engines in ways that emphasize causally incompatible abstractions.

Finally, these abstracted higher-order patterns could support relearning; that is, when individuals encounter an entity again, they have a sense of what are the core phenomena to master. Anecdotally, this seems to be exactly what happens when college students take more advanced courses that build on less advanced survey courses. For example, despite having forgotten many of the details that they were taught in an introduction to biology course, when taking a more advanced class, for instance, in genetics, students may nevertheless show some cognitive savings. For example, they might have a better sense of what to focus on in order to relearn earlier knowledge and to acquire new information in the same topic area.

This list is not meant to be exhaustive, but it illustrates a diverse array of benefits from exposure to mechanisms. We suspect there are many more benefits to having these mechanistically derived abstractions.

Some consequences of knowledge outsourcing

We have argued that exposure to mechanisms enables the induction of higher-order abstractions that endure in memory even as mechanistic details fade. Many of the roles of such abstractions can be viewed as supporting the outsourcing of knowledge. As noted earlier, in our complexity studies, we documented a close relationship between perceived mechanistic complexity and the perceived need for consulting experts (Kominsky et al., 2018). Outsourcing is so natural and often so seamless, even for young children, that it may lead to illusions of much greater self-acquired knowledge than really exists. That is, where individuals routinely rely on a rich web of knowledge interdependencies—on knowledge stored in other minds—people may systematically overestimate the extent of their own individual knowledge because the outsourcing process becomes so automatic and ubiquitous. For example, if one knows exactly whom to consult to answer questions about how refrigerators work, one "knows" how refrigerators work by proxy; however, one may mistake the confidence of knowing the information is available with having it in one's own mind.

We have documented this knowledge-outsourcing effect in a very different paradigm related to internet search (Fisher, Goddu, and Keil, 2015). We conjectured that people become so used to accessing information through search engines that they might confuse knowledge that is in their heads with knowledge that is just a few keystrokes away. This intuition gains credence when one considers feelings of knowledge helplessness that can emerge when one is suddenly and involuntarily disconnected from the internet, such as during an electrical outage. To explore this idea experimentally, we had people search for information on the internet and then compared their intuitions about knowledge with those who were provided the same information in a printed format. We found that simply engaging an internet search induces a greater sense of knowledge mastery for searchable information unrelated to the previously searched information. This is in contrast to others who acquire this information without search. It is important to mention that people do not overestimate their understanding of phenomena that they see as intrinsically unsearchable (such as, why one's parents chose their respective careers).

Here, we argue that an analogous process happens when one chooses causal abstractions to navigate the division of cognitive labor. One tends to overestimate how much one acquired information on one's own, as opposed to how much one needed to rely on other minds. This misplaced sense of knowledge ownership may then contribute to the IOED. One thinks one understands phenomena in more detail than one really does because one can be so successful at acquiring information from others when needed.

These illusions of having more knowledge in one's own mind than one really does can clearly lead to judgment errors, but they also may have an adaptive role. In particular, they might guide people to not spend too much time trying to master and remember details when it can be so easily accessed in other minds when needed. It may be a misplaced sense of knowing, but it is tracking access to knowledge in ways that might be quite reliable. We have shown an analogous effect with respect to intuitions about one's ability to distinguish word meanings, such as those that distinguish a ferret from a weasel or a fir from a pine (Kominsky and Keil, 2014). When people have strong beliefs about other experts easily being able to distinguish between two word meanings, they often mistakenly assume they can make such distinctions on their own. They don't make such mistakes as often for cases where experts do not know either. Illusions of knowing may result in a kind of cognitive economy that reduces storage loads (Wilson and Keil, 1998).

A related phenomena, known as "the community of knowledge effect," serves to illustrate the pervasive nature of outsourced knowledge (Sloman and Rabb, 2016). When participants are told that others understand a phenomenon very well, they are inclined to think that they themselves understand the phenomena better than when they think others do not understand it well. It appears that simply knowing extensive knowledge exists somewhere in the community causes some illusory "leakage" of that knowledge into one's own mind. Although it has not been investigated specifically, it seems likely that the community of knowledge effect may be strongest when that knowledge is described as mechanistic understanding. Simple knowledge of facts or less coherent causal properties would likely not create as much knowledge "contagion."

Remaining questions

Research on mechanistic understanding is still in the early stages, and many questions remain, a number of which may be relevant to philosophy of science.

What is the nature of the memory traces that represent abstract causal information? Is there a particular level of analysis that is privileged? It is clear that fully detailed clockwork images are too fine-grained, but it is less obvious what is the optimal level at which more enduring representations are stored. It is also unclear to what extent these abstractions can be implicit. Some preverbal infants may acquire abstractions based on mechanistic exposure, and if so, are those abstractions largely implicit? At the other end of the continuum, to what extent do practicing scientists also learn without explicit awareness of abstractions that they have derived from exposure to mechanisms?

What kinds of mechanistic exposure optimally lead one to induce these causal abstractions? There clearly seems to be a sweet spot in terms of detail, but its nature needs to be better understood. How much of an idealization of an actual mechanism is desirable? If one is trying to understand how an internal combustion engine works, it seems inappropriate to delve into the details of thermodynamics, even if such details are ultimately essential to a full understanding of the mechanism. A mechanistic account needs to factor out minor or unreliable causal influences as well (Strevens, 2008). What kinds of distortions from the real-world truth are actually better for learning and as a platform for making inductions?

In a related vein, we need to better understand how explanatory goals and other framing effects influence the optimal level of mechanisms that is needed (see also Vasilyeva, Wilkenfeld, and Lombrozo, 2017). The same entity or phenomenon might be embedded in several distinct causal networks that invoke different kinds of explanations. For example, economic considerations and goals might lead to a very different explanation of an artifact's properties than mechanical functional considerations and goals. Even in the same causal network, the level of detail may vary as a function of goals, such as making precise predictions, understanding boundary conditions, and assessing durability and reliability.

We need to more carefully specify what is distinctive about mechanistic understanding. To what extent is it important to have functional properties attached to each of the subcomponents between which there are interactions? How different are teleological explanations from mechanistic ones? (See, e.g., Lombrozo and Wilkenfeld, in press.) In what contexts do preferences for teleological explanations trump others (e.g., Keleman, 1999, 2003), and when is a mechanism preferred? Does mechanism mean different things across different domains such as biology, engineered artifacts, and nonliving natural kinds? To what extent is hierarchical structure important to laypeople in terms of the attraction of mechanisms? Many real-life causal systems involve extensive feedback loops that disrupt simple hierarchies (Bechtel, in press). Similarly, many natural kinds cohere as staple entities because of an intricate web causal homeostasis (Boyd, 2010). Are mechanistic explanations of these other types also cognitively appealing or are there distortions toward more canonical hierarchical structures? Do such distortions persist in practicing scientists?

There are reasons to think that these distortions might endure, given other persistent influences of earlier ways of understanding (Shtulman, 2017). For example, even adults with extensive science backgrounds, when engaged in a time-constrained reasoning task, show differential delays when verifying two types of statements as quickly as they can. For one set of statements, the truth values were the same across both folk and scientific accounts of the event described (e.g., "The moon revolves around the Earth"). For the other set of statements, very similar relations were involved, but the truth values were not the same between folk and formal science versions (e.g., "The Earth revolves around the sun"). All participants verified the cases with conflicting truth values more slowly and with more errors than those with compatible truth values (Shtulman and Valcarcel, 2012). Results like these suggest that early cognitive distortions related to mechanistic understanding may also endure even in the minds of scientists.

These are just a few lines of future inquiry relevant to the distinctive role of mechanistic understanding.

Conclusions

Philosophers of science have come to embrace the importance of a mechanistic way of understanding and doing science. Scientists, especially in the biological sciences, often make progress by uncovering and articulating mechanisms at multiple levels of analysis. Mechanistic explanations also play an important role in folk science, namely, in how laypeople make sense of the world around them. Laypeople consistently seek out mechanisms in their attempts to make sense of various phenomena. Indeed, even preschool children show a consistent preference for mechanistic explanations over other responses to "why" and "how" questions.

Yet, this behavior is also puzzling. Most adults seem to have almost complete ignorance of the specific mechanisms underlying everyday devices and biological systems. Young children are even worse. Why, then, is mechanism an attractive line of inquiry even before children start formal schooling?

The answer seems to require a new account of the role of mechanisms in gaining scientific understanding. Children, as well as all older individuals, derive benefits from exposure to mechanistic explanations even as research repeatedly shows that knowledge of the details soon evaporates in memory. Instead, mechanisms provide a critical platform for making inductions about higher-order abstract causal patterns. In contrast to clockwork details, these high-level abstractions do persist in memory.

These patterns contribute to powerful illusions of explanatory depth. One way is by confusing levels of analysis, wherein high-level functional insights and abstractions are confused with knowing mechanistic details. Those confusions may occur because those higher-order patterns enable even young children to navigate the division of cognitive labor that is central to both the folk sciences and the formal sciences. Because those patterns can enable seamless access to mechanistic details in other minds, they can create a misplaced sense that such knowledge is in one's own mind. In such cases, the IOED may often be an adaptive illusion about accessibility of knowledge. That is, it reflects " knowing" in an indirect sense and may help people from being swamped by attempts to store details that can be accessed relatively easily when needed.

Major questions remain in terms of specifying in detail the traces that are left behind after mechanistic details fade and in experimentally showing the many cognitive roles those patterns play in such activities as relearning, evaluating explanations, and diagnosing problems in real time. In turn, this reconceptualization of the role of mechanisms in folk science may suggest a different view of the role of mechanisms in formal science as well.

References

Ahl, R. E. (2018). *The Complications of Complexity: How Do Developing Minds Represent Complex Artifacts?* Doctoral Dissertation, Yale University.

Ali, A., Veeranki, S. N., Chinchole, A., and Tyagi, S. (2017). MLL/WDR5 complex regulates Kif2A localization to ensure chromosome congression and proper spindle assembly during mitosis. *Developmental Cell*, 41(6): 605–22.

Alter, A. L., Oppenheimer, D. M., and Zemla, J. C. (2010). Missing the trees for the forest: A construal level account of the illusion of explanatory depth. *Journal of Personality and Social Psychology*, 99(3): 436.

Baer, C. and Friedman, O. (2017). Fitting the message to the listener: Children selectively mention general and specific facts. *Child Development*. Doi: 10.1111/cdev.12751

Bechtel, W. (2011). Mechanism and biological explanation. *Philosophy of Science*, 78: 533–57. Doi: 10.1086/661513

Bechtel, W. (in press). *The Importance of Constraints and Control in Biological Mechanisms: Insights from Cancer Research*. Philosophy of Science.

Bechtel, W. (2017a). Using the hierarchy of biological ontologies to identify mechanisms in flat networks. *Biology and Philosophy*, 32: 627–49.

Bechtel, W. (2017b). Top-down causation in biology and neuroscience: Control hierarchies. In F. Orilia and M. P. Paoletti (Eds.), *Philosophical and Scientific Perspectives on Downward Causation*. London: Routledge.

Bechtel, W. and Richardson, R. C. (1993). *Discovering Complexity: Decomposition and Localization as Strategies in Scientific Research*. Cambridge: MIT Press.

Bechtel, W. and Richardson, R. C. (2010). *Discovering Complexity: Decomposition and Localization as Strategies in Scientific Research*. MIT Press.

Boyd, R. (2010). Homeostasis, higher taxa, and monophyly. *Philosophy of Science*, 77(5): 686–701.

Canfield, J. (1964). Teleological explanation in biology. *The British Journal for the Philosophy of Science*, 14(56): 285–95.

Chouinard, M. M., Harris, P. L., and Maratsos, M. P. (2007). Children's questions: A mechanism for cognitive development. *Monographs of the Society for Research in Child Development*, 72: i–129.

Cimpian, A. and Markman, E. M. (2009). Information learned from generic language becomes central to children's biological concepts: Evidence from their open-ended explanations. *Cognition*, 113(1): 14–25. Doi: 10.1016/j.cognition.2009.07.004

Cimpian, A. and Petro, G. (2014). Building theory-based concepts: Four-year-olds preferentially seek explanations for features of kinds. *Cognition*, 131(2): 300–10. Doi: 10.1016/j.cognition.2014.01.008

Craver, C. and Kaplan, D. M. (2018). Are more details better? On the norms of completeness for mechanistic explanations. *The British Journal for the Philosophy of Science.*

Dolnick, E. (2011). *The Clockwork Universe: Isaac Newton, Royal Society, and the Birth of the Modern World I.* New York: Harper Collins.

Faber, R. J. (1986). *Clockwork Garden: On the Mechanistic Reduction of Living Things.* University of Massachusetts Press. Doi: 10.5840/schoolman198966341

Felline, L. (2018). Mechanisms meet structural explanation. *Synthese*, 195(1): 99–114.

Fisher, M. and Keil, F. C. (2016). The curse of expertise: When more knowledge leads to miscalibrated explanatory insight. *Cognitive Science*, 40: 1251–69.

Fisher, M., Goddu, M. K., and Keil, F. C. (2015). Searching for explanations: How the Internet inflates estimates of internal knowledge. *Journal of Experimental Psychology: General*, 144(3): 674.

Fisher, M. and Keil, F. C. (in prep). Decay neglect: An illusion of knowledge persistence in students.

Frazier, B. N., Gelman, S. A., and Wellman, H. M. (2009). Preschoolers' search for explanatory information within adult–child conversation. *Child Development*, 80(6): 1592–611.

Frazier, B. N., Gelman, S. A., and Wellman, H. M. (2016). Young children prefer and remember satisfying explanations. *Journal of Cognition and Development*, 17(5): 718–36. Doi: 10.1080/15248372.2015.1098649

Frisch, M. (2014). *Causal Reasoning in Physics.* Cambridge: Cambridge University Press. Doi: 10.1017/cbo9781139381772

Glennan, S. (2017). *The New Mechanical Philosophy.* Oxford: Oxford University Press.

Glennan, S. and Illari, P. (2018). *The Routledge Handbook of Mechanisms and Mechanical Philosophy.* Oxford: Routledge.

Hickling, A. K. and Wellman, H. M. (2001). The emergence of children's causal explanations and theories: Evidence from everyday conversation. *Developmental Psychology*, 37: 668–83. Doi:10.1037/0012-1649.37.5.668

Hussak, L. J. and Cimpian, A. (2015). An early-emerging explanatory heuristic promotes support for the status quo. *Journal of Personality and Social Psychology*, 109(5): 739–52.

Johnson, S. G. and Keil, F. C. (2014). Causal inference and the hierarchical structure of experience. *Journal of Experimental Psychology: General*, 143(6): 2223.

Johnston, A. M., Sheskin, M., Johnson, S. G., and Keil, F. C. (2018). Preferences for explanation generality develop early in biology but not physics. *Child Development*, 89(4): 1110–19.

Kästner, L. and Andersen, L. M. (2018). Intervening into mechanisms: Prospects and challenges. *Philosophy Compass*, e12546.

Keil, F. C. (2006). Explanation and understanding. *The Annual Review of Psychology*, 57: 227–54.

Keil, F. C., Stein, C., Webb, L., Billings, V. D., and Rozenblit, L. (2008). Discerning the division of cognitive labor: An emerging understanding of how knowledge is clustered in other minds. *Cognitive Science*, 32(2): 259–300. Doi: 10.1080/03640210701863339

Kelemen, D. (1999). The scope of teleological thinking in preschool children. *Cognition*, 70(3): 241–72. Doi: 10.1016/s0010-0277(99)00010-4

Kelemen, D. (2003). British and American children's preferences for teleo-functional explanations of the natural world. *Cognition*, 88(2): 201–21. Doi: 10.1016/S0010-0277(03)00024-6

Kominsky, J. F. and Keil, F. C. (2014). Overestimation of knowledge about word meanings: The "misplaced meaning" effect. *Cognitive Science*, 38(8): 1604–33.

Kominsky, J. F., Zamm, A. P., and Keil, F. C. (2018). Knowing when help is needed: A developing sense of causal complexity. *Cognitive Science*, 42(2): 491–523.

Kurkul, K. E. and Corriveau, K. H. (2017). Question, explanation, follow-up: A mechanism for learning from others? *Child Development*. Doi: 10.1111/cdev.12726

Kushnir, T. and Gelman, S. A. (2016). Translating testimonial claims into evidence for category-based induction. A. Papafragou, D. Grodner, D. Mirman, and J. C. Trueswell (Eds.), *Proceedings of the 38th Annual Conference of the Cognitive Science Society*. Austin: Cognitive Science Society.

Lawson, R. (2006). The science of cycology: Failures to understand how everyday objects work. *Memory and Cognition*, 34(8): 1667–75.

Lecompte, G. K. and Gratch, G. (1972). Violation of a rule as a method of diagnosing infants' levels of object concept. *Child Development*, 43: 385–96.

Lockhart, K. L., Chuey, A., Kerr, S., and Keil, F. C. (under review). *The Privileged Status of Knowing Mechanistic Information: An Early Epistemic Bias*. https://doi.org/10.1111/cdev.1324

Lockhart, K. L., Goddu, M. K., Smith, E., and Keil, F. C. (2016). What could you learn on your own?: Understanding the epistemic limitations of knowledge acquisition. *Child Development*, 87: 477–93. Doi:10.1111/cdev.12469

Lombrozo, T. (2007). Simplicity and probability in causal explanation. *Cognitive Psychology*, 55: 232–57.

Lombrozo, T. and Wilkenfeld, D. (in press). Mechanistic versus functional understanding. In Stephen R. Grimm (Ed.), *Varieties of Understanding: New Perspectives from Philosophy, Psychology, and Theology*. New York: Oxford University Press.

Machamer, P. K., Darden, L., and Craver, C. F. (2000). Thinking about mechanisms, *Philosophy of Science*, 67: 1–25.

Mills, C. M. and Keil, F. C. (2004). Knowing the limits of one's understanding: The development of an awareness of an illusion of explanatory depth. *Journal of Experimental Child Psychology*, 87(1): 1–32.

Roe, S. M. and Baumgaertner, B. (2017). Extended mechanistic explanations: Expanding the current mechanistic conception to include more complex biological systems. *Journal for General Philosophy of Science*, 48(4): 517–34.

Rosenberg, A. (2018). Making mechanism interesting. *Synthese*, 195(1): 11–33.

Rozenblit, L. R. and Keil, F. C. (2002). The misunderstood limits of folk science: An illusion of explanatory depth. Cognitive Science, *26*, 521–562.

Russell, B. (1912, January). On the notion of cause. In *Proceedings of the Aristotelian society* (Vol. 13, pp. 1–26). Aristotelian Society, Wiley.

Shtulman, A. (2017). *Scienceblind: Why Our Intuitive Theories about the World Are So Often Wrong*. New York: Hachette UK.

Shtulman, A. and Valcarcel, J. (2012). Scientific knowledge suppresses but does not supplant earlier intuitions. *Cognition*, 124(2): 209–15.

Sloman, S. A. and Rabb, N. (2016). Your understanding is my understanding: Evidence for a community of knowledge. *Psychological Science*, 27(11): 1451–60.

Strevens, M. (2008). *Depth: An Account of Scientific Explanation*. Cambridge: Harvard University Press.

Strickland, B., Silver, I., and Keil, F. C. (2017). The texture of causal construals: Domain-specific biases shape causal inferences from discourse. *Memory & Cognition*, 45(3): 442–55.

Trouche, E., Chuey, A., Lockhart, K. L., and Keil, F. C. (2017). Why Teach How Things Work? Tracking the Evolution of Children's Intuitions about Complexity. In *Proceedings of the 39th Annual Conference of the Cognitive Science Society*. (pp. 3368–3373) Austin, TX: Cognitive Science Society.

Vasilyeva, N., Wilkenfeld, D., and Lombrozo, T. (2017). Contextual utility affects the perceived quality of explanations. *Psychonomic Bulletin & Review*, 1436–50. Doi: 10.3758/s13423-017-1275-y

Wilson, R. A. and Keil, F. C. (1998). The shadows and shallows of explanation. *Minds and Machines*, 8: 137–59. Reprinted in revised form: R. A. Wilson and F. C. Keil (2000). In F. C. Keil and R. A. Wilson (Eds.), *Explanation and Cognition*. Cambridge: MIT Press.

Wright, L. (1976). *Teleological Explanations: An Etiological Analysis of Goals and Functions*. Berkeley: University of California Press.

Part Two

Theories and Theory Change

4

Information That Boosts Normative Global Warming Acceptance without Polarization: Toward J. S. Mill's Political Ethology of National Character

Michael Andrew Ranney, Matthew Shonman, Kyle Fricke,
Lee Nevo Lamprey, and Paras Kumar

Introduction

The history of psychology (e.g., Sahakian, 1968) is marbled with debates, of varying scales, about the relative degrees to which top-down or bottom-up processes dominate particular mental phenomena and performances. In this chapter, we consider this theory-driven/data-driven dialectic with respect to (1) four new, pertinent, experiments in which factual material increases people's acceptance that global climate change is occurring/concerning, and (2) conceptually adjacent philosophical considerations related to rationalism and national identity. The experiments collectively demonstrate (occasionally replicating) five brief interventions that boost people's acceptance of (e.g., anthropogenic) global warming (GW)—without yielding polarization. Experiment 1 demonstrates GW acceptance gains across ten mini-interventions contrasting graphs of Earth's temperature rise and equities' valuations rise. Experiment 2 replicates and extends earlier work showing GW acceptance changes among participants receiving feedback for a handful of their estimates regarding climate statistics. Experiment 3 exhibits a GW acceptance gain among US participants receiving feedback for nine of their estimates regarding "supra-nationalism" statistics that suppress American (over-)nationalism. Experiment 4's large contrastive study compares Experiment 2's representative stimuli's effect with (a) the effect sizes of five videos and three texts that differentially explain global warming's mechanism, as well as (b) non-mechanistic controls; Experiment 4 additionally serves to further replicate some (textual and statistical) findings from prior research and bolster Experiment 3's findings. The interventions' success further justifies John Stuart Mill's and others' perspective that humans often function as empiricists. (Many of the interventions are—fully or partially—available at www.HowGlobalWarmingWorks.org, our multi-language website for directly enhancing all nations' public "climate change cognition.")

The reasoning researcher James F. Voss once told the first author that (paraphrasing) "psychology's history suggests that, any time a cognitive scholar considers explanatorily excluding one of two major contrastive yin-yang possibilities—such as either top-down or bottom-up processing—s/he is likely to ultimately be seen as mistaken." This chapter's experiments' findings, we believe, simultaneously (a) provide some insight into Americans' often-peculiar thinking with respect to global warming, and (b) cohere, in this realm, with Voss's synergistic generalization (in this incarnation: that individuals can change their minds due to either, or both, evidence-based and culture-based cognition—although we empirically focus mostly on the former herein). We will present our four new experiments about GW cognition in the spirit of J. S. Mill's proposed "political ethology"—a science of national character. (Domains attendant to GW cognition include, among others we have identified, the psychology of evolution, religion, nationalism, affiliation, and conservatism; e.g., Ranney and Thanukos, 2011.) We start here by considering Mill's musings about how nations might be differentiated—in relief to the much more recent Reinforced Theistic Manifest Destiny theory (RTMD; e.g., Ranney, 2012).

Some critical ways in which nations might differ: RTMD as a Millian Theory

International polls are increasingly salient and frequently illuminating, such as recent polling about how nations' populaces view Vladimir Putin (sometimes in comparison with Donald Trump; Vice/Pew, 2017). John Stuart Mill (1843/2006) may not have imagined such samplings when considering "a State of Society" (p. 911) or "the character of the human race" more broadly (p. 914), but his philosophy entertained both individual and sociological mental analyses in articulating his senses of the differential characters of nations—while highlighting "psychological and ethological laws" on the development of "states of the human mind and of human society" (p. 914). Mill envisioned a science that would "ascertain the requisites of stable political union" (p. 920), a stability that he ascribed to England, among some other nations (but certainly not all[1]). One of Mill's essential conditions of a stable political society is a sense of "nationality" (which he contrasts to its more negative senses)—a citizenry's "feeling of common interest" that provides "cohesion among the members of the same community or state" (p. 923). The experiments presented below are part of a scientific research program that (with Ranney and Clark, 2016, etc.), in various ways, probes aspects of US national character (Ranney, 2012; Ranney and Thanukos, 2011)—thus implementing the kind of science Mill seemed to advocate.

Recognizing political ethology, Mill called for a science of national character. In modern terms, he seemed to effectively posit the existence of certain local cognitive minima in the energy-space of national identities—in other words, that there are essentially discrete, stable configurations of societal positions about important issues. Rather as behaviorists noted that not all stimuli and potential consequences are equally naturally associable (e.g., Garcia and Koelling, 1966), on a molar level, the associationist (Sahakian, 1968) Mill argued: "When states of society, and the causes which produce them, are spoken of as a subject of science, it is implied that there exists a natural correlation among these different elements; that not every variety of combination of these general social facts is possible, but only certain combinations; that, in short, there exist Uniformities of Coexistence between the states of the various social phenomena" (Mill, 1843/2006, p. 912).

Earlier in this decade, the RTMD theory was induced (Ranney, 2012; Ranney and Thanukos, 2011) to explain some of the kinds of inter-"element" correlations that Mill might consider, were he researching today. RTMD may best be considered quasi-causal in that, rather as Mill might have noted of nineteenth-century England, some RTMD relationships approximate a logically competitive character (e.g., at an extreme today: if biblical creation is thought to be literally true, that person sees evolution as much less plausible because they compete). In contrast, other relationships approximate historical concept-affiliations (e.g., fossil-fuel-poor countries, *ceteris paribus*, generally have fewer nationalist-energy reasons to deny climate change [setting aside those countries' development-inhibition fears]). Figure 4.1 illustrates RTMD (which is more extensively explicated elsewhere, especially in Ranney, 2012), which embodies relatively stable national relationships among six main constructs—both noticed and predicted—including acceptance/affinity regarding (a) an afterlife, (b) a deity (/deities), (c) creationism, (d) nationalism, (e) evolution, and (f) global warming (GW; e.g., the focal aspects of climate change—that Earth is generally warming, largely anthropogenically, and therefore warrants considerable concern). For Americans, and presumably for *any* nation's people, one's acceptance of (a)–(d) are predicted to correlate with each other (e.g., caricaturing the religious

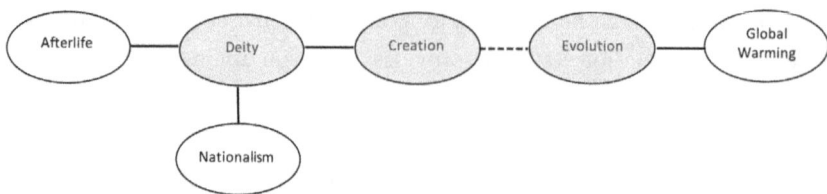

Figure 4.1 Reinforced Theistic Manifest Destiny (RTMD) theory (Ranney et al., 2012a), which extended a more "received view" perspective (i.e., the three shaded ovals) regarding some classically controversial socio-scientific constructs. A geopolitical theory, RTMD posits coherent or conflicting notions with, respectively, solid or dashed conceptual links. Especially for a given individual (and plausibly for any given nation), RTMD predicts that the degree to which the four leftmost constructs are accepted, the less the rightmost two constructs will be accepted, and vice versa. Correlational and causal data to date have overwhelmingly borne this out.

right: "God and country," "joining God in heaven," "God created organisms as they are," etc.), as are (e)–(f) (e.g., caricaturing the less religious left: "evolution has, through humans, caused global warming"). However, each construct of {a–d} should negatively correlate (i.e., anticorrelate) with each construct of {e–f}, a prediction that is causally demonstrated in Experiments 3 and 4 (e.g., d-versus-f: "solving global warming involves *inter*national agreements and less über-nationalism"). Figure 4.1 highlights the three central (theistically impacted) constructs from RTMD's original explication—deity, creation, and (creation's strongly anticorrelated associate) evolution—from which much of nations' conceptual tensions may stem. (See also Figure 4.1's caption.)

With other colleagues from Berkeley's Reasoning Group, we have assessed these six constructs' interrelationships using various US participant populations, and when sample sizes permit, the fifteen correlations among (a)–(f) always appear in the predicted directions, usually with statistical significance (e.g., Ranney et al., 2012a). Indeed, we have never found a significant US correlation opposite in direction to that which RTMD theory predicts (although for some rather small-sample studies focused on "boutique" hypotheses, we occasionally find *non*significant correlations in contrapredicted directions).

People are empiricists (and rationalists)

RTMD theory has inspired many GW-focused experiments (e.g., Ranney and Clark, 2016; Ranney, Munnich, and Lamprey, 2016), and four more are explicated herein. Beyond more specific hypotheses, each experiment assesses whether people are empiricists—in the sense that they will change their views to better cohere with presented scientific (e.g., statistical or empirically induced mechanistic) information. We show below, as previously, that they do. These findings hardly negate rationalistic (e.g., theory-conserving) processes, and the findings hardly suggest that people are merely bottom-up thinkers—because top-down, hypothesis-driven rumination is assumed to *also* be fundamental to human cognition (e.g., Ranney, 1987; cf. Sahakian, 1968, p. 2, regarding Anaxagoras, Democritus, and reason vs. the senses).[2]

Many, like we, believe psychologists had *already* proven that humans are empiricists in that they change beliefs in the face of empirical evidence (e.g., van der Linden et al., 2017). However, the occasional researcher claims that cultural biases eliminate the possibility of changing people's minds with scientific information (e.g., Kahan et al., 2012). This position, which Kahan and his colleagues ascribe to people, is so extreme that no noteworthy philosopher appears to have held it[3]—and it might be seen as an untenably extreme (e.g., fully top-down) form of some rationalists' *superiority of reason thesis*[4] (Markie, 2017), which would also conflict with the data priority principle of explanatory coherence theory (Ranney and Schank, 1998; Ranney and Thagard, 1988; Thagard, 1989, etc.). Kahan and Carpenter (2017) continued this overly rationalist attribution to people, inaccurately[5] asserting an "immobility of public opinion" on GW, even as they noted that "decision scientists have been furiously" trying to improve "public engagement with this threat" (p. 309). Ranney and Clark (2016) extensively disconfirmed this (essentially anti-empiricist) "stasis" view through controlled experiments, historical counterexamples, and a literature review of crucial research (see, e.g., Ranney & Clark's pp. 50–55 and 65–67) showing that even Kahan et al.'s own (2015) data disconfirmed the stasis view (also see Ranney, Munnich, and Lamprey, 2016). More recently, van der Linden et al. (2017) have more briefly elaborated upon Ranney and Clark's (2016) exposition that culture versus information is a false dichotomy, calling "culture versus cognition" a "false dilemma."

Common sense and Mill's work both share our predilection toward empirical information as crucial in determining beliefs. At a gut-phenomenal level, people are strikingly changed by even a single observation (also see Ranney, Munnich, and Lamprey, 2016). Examples include (a) watching a would-be recommendation letter-writer fatally flattened by a road roller—or (b) returning home unexpectedly to observe a supposedly monogamous partner engaged in infidelity. One could reject such bottom-up evidence through some *extreme* reason-superiority rationalist interpretation, but that rarely happens. In another contrastive example, (c) people ending their vacations to find their house burned/shaken/blown/washed to its foundation don't just deny it, as they see that their structure is gone (and denial might warrant psychiatric hospitalization). The datum, in the moment, changes one forever. We are empiricists, even if we are also—perhaps equally or more at times—top-down rationalists with *non*extreme intransigence and/or *non*extreme affections regarding extant beliefs. (Naturally, a single rational inference also, at times, transforms what seems empirically settled, e.g., regarding explanatory coherence: Ranney and Thagard, 1988; Ranney and Schank, 1998.) The empiricist Mill (1843/2006) highlighted new knowledge's crucial role in changing cultures/societies (just as we have highlighted mind-changing information: Ranney and Clark, 2016; Ranney, Munnich, and Lamprey, 2016). Although celebrating the "progress of knowledge" and lamenting that "the changes in the opinions" of people are slow, Mill noted that every "considerable advance in material civilization has been preceded by an advance in knowledge" (p. 927).

Global climate change denial

Early in *this* century, many characterize the most frustratingly slow opinion change to be Americans'—and many of their representatives'—reluctance to quickly inhibit global climate change. Mill induced overall societal progress, yet saw nations *nonmonotonically* improving—seemingly anticipating *local* optimization notions (cf. Ranney and Thagard, 1988; Thagard, 1989, etc.): "the general tendency is, and will continue to be, saving occasional and temporary exceptions, one of improvement; a tendency towards a better and happier state" (1843/2006, p. 913, etc.). Our materials and our website, HowGlobalWarmingWorks.org (Ranney and Lamprey, 2013–present), discussed below, use scientific information to foster such improvements regarding climate change beliefs—hopefully even "flipping" beliefs. Our efforts are meant to better engage Mill's "social science," of which he said (p. 912) the "fundamental problem . . . is to find the laws according to which any state of society produces the state which succeeds it and takes its place." In modest respects, RTMD represents seeking such laws (e.g., in positing and confirming an inverse nationalism–GW relationship).

Every nation signed the Paris climate change accord, yet America announced plans to leave it. US acceptance of the following propositions lag behind its peer nations on an alarm-spectrum of beliefs, including that GW is (a) occurring, (b) partially anthropogenic, (c) largely anthropogenic,

(d) imminently concerning, and (e) demanding fast action. So, although RTMD ought to apply to virtually all nations, we focus our instruction on participants from the United States, where urgent need seems greatest, especially given that America presently represents the largest economic, military, and (arguably) political force in the world. Therefore, this chapter's four new experiments each further assesses the hypothesis that—and the degree to which—empirical information can increase Americans' GW acceptance. Extending our prior efforts, Experiment 1 assesses a new dimension of potentially persuasive empirical information: contrastive graphs. Experiment 2 both extends and replicates a prior finding (from Ranney and Clark, 2016, whose seven experiments mostly focused on the utility of explaining GW's mechanism) regarding the effectiveness of *statistics* to increase GW acceptance—with improved stimuli and a larger sample. Experiment 3 assesses a novel RTMD prediction (related to Mill's interest in studying national character): that reducing one's super-nationalism with supra-nationalist statistics will increase the person's GW acceptance. Experiment 4 represents a twenty-one-condition mega-experiment that contrasts the usefulness of many interventions to consider their relative "bang per buck" in increasing such acceptance. In essence, Experiments 1 and 2 address the stasis—or "science impervious"—view again (re: empiricism), Experiment 3 engages RTMD theory, and Experiment 4 does both.

Experiment 1: Boosting GW acceptance with graphs and averaging

Building on prior research showing that statistical and mechanistic information (separately) increase GW acceptance (Ranney and Clark, 2016; Ranney, Munnich, and Lamprey, 2016), we hypothesized that visual graphs representing observed temporal data trends could similarly effectively and durably increase such acceptance. Utilizing graphs of Earth's average surface temperature and the Dow Jones Industrial Average (adjusted for inflation: "DJIA-a"), Experiment 1 participants responded to queries encouraging them to reflectively reason and critically process the graphs' information.

This experiment (Chang, 2015; cf. Ranney, Chang, and Lamprey, 2016) was inspired by Lewandowsky (2011), who asked participants to extrapolate from a temperature graph or an identical pseudo-stock graph; overall, Lewandowsky's participants extrapolated rises in both global temperature and (purported) stock prices for future years, establishing that graphs depicting temporal temperature trends could increase GW acceptance. Experiment 1 replicated and elaborated Lewandowsky's finding by (a) soliciting broader responses about provided graphs, (b) depicting *actual* stock data (i.e., the DJIA-a), and (c) assessing the longevity of participants' changes with a delayed ("Phase 2") experimental session.

Method

Participants

Participants were 712 workers recruited from Amazon's Mechanical Turk (MTurk). Forty-nine participants were excluded after the experiment's initial phase, and 429 of the retained 663 participants both chose to complete, and satisfactorily completed, the experiment's second phase—a delayed posttest that was administered after 6–9 (M = 8.6) days. Participants were excluded if they (a) were not US citizens located in the United States, (b) took excessive or too little time relative to their peers, (c) failed too many attention checks, and (d) generated *overly* incoherent/self-contradictory data, as determined through index scores. An example of (d) would be a participant fully accepting a deity's existence on one item, yet fully rejecting any deities' existence on a later item (which suggests rushed, "random," or inattentive responding).

Design, procedure, and materials

All the experiments' materials were presented using Qualtrics. Experiment 1's participants responded, using a 1–9 Likert scale[6], to the same set of eight GW items during three testing times: twice in Phase 1—during pretesting and immediate posttesting—and once in Phase 2's delayed posttest (9 days hence). Table 4.1 presents the items, which were similar to GW items from Ranney and Clark (2016) and so on. Following pretesting, participants learned about an unbiased alien-robot, "Bex," who helped guide them through one of ten randomly assigned conditions and helped them understand the data and the utility of averaging when analyzing graphical data over long periods. (Chang, 2015, explicated the conditions, which were specifically selected from 48 [3 × 2 × 2 × 2 × 2 factorial] possibilities to test particular hypotheses.)

Participants rated the degree to which they agreed with each statement, selecting a number from 1 (Extremely Disagree) to 9 (Extremely Agree).

In each condition, participants received graphs concerning the Earth's surface temperature and the DJIA-a. The Earth's air and surface temperature data,

Table 4.1 Global warming attitude items on pre- and posttest. (Participants rated the degree to which they agreed with each statement, selecting a number from 1 [Extremely Disagree] to 9 [Extremely Agree].)

Item Text	Response Range
Human activities are largely responsible for the climate change (global warming) that is going on now.	1–9
Global warmings or climate changes, whether historical or happening now, are only parts of a natural cycle.	1–9
If people burned all the remaining oil and coal on Earth, the Earth wouldn't be any warmer than it is today.	1–9
I am confident that human-caused global warming is taking place.	1–9
I am concerned about the effects of human-caused global warming.	1–9
I would be willing to vote for a politician who believes human-caused global warming doesn't occur.	1–9
Global warming (or climate change) isn't a significant threat to life on Earth.	1–9
The Earth isn't any warmer than it was 200 years ago.	1–9

spanning 1880–2014, were from NASA. The DJIA-a data reflected US financial stock/equity values during 1885–2014 (Williamson, 2015). Because *annually* averaged scatter-plots increase the difficulty of honing in on temperature and DJIA-a trends, beyond-annual averaged graphs were provided that used two forms of averages: moving and span. With these averages, we created DJIA-a and temperature graphs that showed 4, 8, 16, and 64-year-averaged trends (pp. 150–151 of Ranney, Munnich, and Lamprey, 2016, displays six of these graphs, and 16-year span-averaged graphs are displayed at howglobalwarming works.org/2graphsab.html). Participants were randomly assigned to one of the ten conditions, which differed in the number and types of graphs shown, and so on. The ten interventions were brief—averaging 6 minutes, with the longest taking 7 minutes. (Chang, 2015, offers much greater detail.)

During the interventions, each participant was asked to complete/answer sets of interactive exercises/questions designed to facilitate numerical reasoning and engagement with the DJIA-a and temperature graphs. Four graphical analysis techniques were utilized. First, participants analyzed whether the averaged stock and/or temperature graphs were increasing, decreasing, or neither. Second, participants differentiated between side-by-side 16-year-averaged DJIA-a and temperature graphs (i.e., to guess/decide which graph was which). Third, participants extrapolated five future data points on both DJIA-a and temperature graphs (in 5-year increments, up to 2035). Finally, participants were told that some participants had received switched graphs (i.e., graphs stated as DJIA-a graphs were temperature graphs and vice versa), and participants were asked whether they believed theirs were switched.

Following the intervention, participants completed an immediate posttest that reassessed GW attitudes and elicited demographic data. Nonexcluded participants were invited to complete a (shorter) second phase 6–9 days after Phase 1. Phase 2's delayed posttest was similar to Phase 1's immediate posttest, in that it again reassessed GW attitudes (to assess the immediate posttest gains' durability), but participants were *not* reintroduced to Bex, the logic of averaging, or why span- and moving-average graphs are useful. Following Phase 2's central portion of the delayed posttest (e.g., focusing on GW beliefs), participants were asked to both (a) differentiate between 16-year-averaged DJIA-a and temperature graphs, and (b) extrapolate DJIA-a and temperature up to 2035.

Results

Participants demonstrated durable, robust, and statistically significant GW acceptance gains for *each* of the ten interventions at *both* immediate and delayed posttests (p values from .0015 to 2.95×10^{-14}). Regarding gains possible (i.e., from pretest acceptance to extreme, "9," acceptance), combining all ten individually significant conditions, the mean gain was roughly 23.4 percent of what was possible on the immediate posttest (.655 points; $t(523) = -20.33$; $p < 2 \times 10^{-16}$) and 20.7 percent on the delayed posttest (.579 points; $t(381) = -16.19$; $p = 2.2 \times 10^{-16}$), demonstrating participants' high retention of the gained acceptance even 9 days postintervention. In addition, 98 percent of participants believed the 16-year span-average temperature graph to be increasing, and the other 2 percent did not believe it decreased. Virtually all participants also predicted that Earth's temperature (and the DJIA-a) would continue to rise through 2035. Furthermore, no polarization was observed; even the most conservative participants (on *both* economic and social subscales) exhibited GW acceptance gains—as did liberals.

Rather as did Garcia de Osuna, Ranney, and Nelson (2004, which regarded abortion-reasoning), we analyzed "flips" and "semi-flips"—here, individuals changing relative to the 5.0 midpoint of a scale. As is typical with our GW intervention experiments, many more participants' average GW scores (a) flipped from below-5.0 to above-5.0 than vice versa (p = .004) or (b) semi-flipped upward (from 5.0 to above-5.0 or from below-5.0 to 5.0) than vice versa (p = .006). Even more strikingly, because considerably fewer participants were below-5.0 on the pretest than above-5.0 on the pretest, it was *12.45 times* more likely (regarding conditional probabilities) that a 5.0-or-lower participant would flip or semi-flip toward greater GW acceptance than that a 5.0-or-higher participant would flip or semi-flip toward lesser GW acceptance.

Discussion

The results show that the graphs' plotted and averaged scientific data (e.g., Earth's surface temperature, and the DJIA-a, over time) help people visualize and sort through noisy data, making trends clear. Our graphs generally become more interpretable when each averaged datum subsumed longer temporal periods (i.e., 4- vs. 8- vs. 16- vs. 64-year-averaged trends). The 64-year moving-average graphs are particularly compelling because their near-monotonicity makes it difficult to deny that the DJIA-a—and, more importantly, Earth's temperature—have been rising since the 1880s.

Our graphs spawn scientific climate change inferences in several ways. First, many people infer the rising temperature trend rather directly after merely viewing the annual (and more aggregated) temperatures since 1880. Second, even if one cannot honestly infer/"see" the temperature rise (for instance, in the 16-year-averaged graphs), if one (a) believes that the DJIA has been increasing, yet (b) cannot confidently differentiate between the DJIA-a and temperature graphs (or if one cannot tell whether the graphs were switched), then one should (c) infer that both equity values and mean temperature are increasing. Additionally, asking participants to extrapolate future annual means encourages participants to analyze past data trends and incorporate them coherently into their knowledge of climate change trends—yielding positively sloping extrapolations.

Experiment 1's findings again (as per Ranney and Clark, 2016; also see Ranney, Munnich, and Lamprey, 2016; and van der Linden et al., 2017) *disconfirm* two suggestions: (a) that conveying germane scientific climate information polarizes conservatives and (b) that effecting conceptual changes through new knowledge is virtually hopeless. The disconfirmed "stasis theory" (i.e., the [b] just mentioned; e.g., see Kahan et al., 2012) underestimates people's abilities to counter purported inclinations toward extreme reason-superiority (i.e., über-rationalist, top-down thinking) and incorporate new, germane, information—such as the temperature data. Experiment 1 replicated prior stasis theory disconfirmations, adding to a growing body of experiments that have successfully used short knowledge-based interventions to change participants' GW acceptance (n.b. Ranney and Clark, 2016, pp. 54–55, provides a partial review; also see Lombardi, Sinatra, and Nussbaum, 2013; Otto and Kaiser, 2014).

No matter which condition participants received, 100 percent of this experiment's ten interventions fostered a significant GW acceptance increase. Strikingly, the observed gains were retained with virtually no decay after 9 days' delay. Given these "ten for ten" conditions' successes in yielding (and overwhelmingly maintaining) GW acceptance gains, this form of intervention seems a useful addition to prior successful instruction (e.g., mechanistic explanations and statistical evidence; Ranney and Clark, 2016, etc.) that have helped convince people that GW is occurring, anthropogenic, and worthy of concern. Each of these ways also reinforces a notion of "human as empiricist" and not just "human as extreme rationalist"—in keeping with Mill's (e.g., 1843/2006) philosophy.

Experiment 2: Changing GW acceptance with representative (and misleading) statistics

Experiment 1 established that brief interventions involving averaged graphs can durably increase one's GW acceptance. Prior studies (Ranney & Clark's Experiments 2–5) found similar durability using mechanistic GW explanations. In Experiment 2, we (Ng, 2015; Ranney, Munnich, and Lamprey, 2016) sought to (a) extend and replicate Ranney and Clark's (2016) Experiment 6, which showed that numerical feedback following participants' estimates of a series of *representative* statistical GW-related quantities also increased GW acceptance, and (b) replicate Ranney and Clark's (2016) Experiment 7 finding that *misleading-albeit-accurate* statistics can induce doubt regarding GW's reality.

Following the Numerically Driven Inference paradigm (NDI; e.g., Ranney et al., 2001; Ranney et al., 2008; Ranney, Munnich, and Lamprey, 2016), which includes inducing learning by providing feedback regarding the veracity of participants' estimates, this experiment also provided the true value of each statistic as feedback. We hypothesized—replicating Ranney and Clark's (2016) Experiments 6 and 7—that feedback on the representative statistics would increase GW acceptance, while providing the *misleading* statistics would *decrease* GW acceptance (e.g., Clark, Ranney, and Felipe, 2013). Extending those two experiments, we sought to assess such changes' durabilities about 9 days later (as in Experiment 1). The null hypothesis is that, perhaps due to efforts by those denying climate change, postintervention effects might be labile (see Ranney, 2008, and Ranney, Munnich, and Lamprey, 2016, on durability); our hope was that, for the representative statistics, increased GW acceptance would be maintained upon delayed posttesting.

Enhancements to Ranney and Clark's (2016) methods included updated statistical feedback values and the expansion of the representative statistics from seven to nine items. Methodology was also improved, with (a) a new example statistic for participants to practice the estimation procedure, (b) instructions better highlighting the feedback statistics as accurate, and (c) added exclusion criteria, such as checks to determine whether participants' answers were self-consistent. This study also incorporated Carol Dweck's "fixed versus growth" mindset concept. Dweck (2006) defines a fixed mindset as regarding "basic qualities, like their intelligence or talents" as "simply fixed traits," and a growth mindset as viewing one's "most basic abilities" as able to "be developed through dedication and hard work." Because climate change challenges people to modify their thinking and behavior, a secondary hypothesis was that the Representative intervention might shift participants, possibly durably, more toward a growth mindset.

Method

Participants

Participants were 282 MTurk workers. Following exclusions, 257 remained (129 receiving representative items and 128 receiving misleading items), and 176 of those (68.48 percent) completed a delayed posttest occurring 6–9 (M = 8.59) days later.

Design, procedure, and materials

Of four (2 × 2) experimental conditions, two received nine representative statistics (exhibited in Ranney, Munnich, and Lamprey, 2016, pp. 158–159) and two received eight misleading statistics (described in Ranney and Clark, 2016). The "Sandwich Representative" and "Sandwich Misleading" groups received both a pretest and immediate posttest; the pretest was omitted for the "Open Representative" and "Open Misleading" groups to assess/counter any effects of perceived experimenter demand. (The Sandwich/Open analogy maps a test to a bread slice; a "sandwich" procedural phase has pretest *and* posttest "slices," whereas the no-pretest procedure here represents an "open" sandwich.) For misleading conditions, Phase 1 (i.e., the pre-delay phase) ended with a debriefing on the facts behind GW, which was meant to reverse any induced beliefs denying climate change.

Participants' attitudes and mindsets were assessed during the pretest (for the Sandwich groups), immediate posttest, and delayed posttest. The RTMD surveys used the nine-point scale described above, while the mindset survey used its six-point scale.

For each statistical quantity/item, participants were prompted to (a) estimate its value, (b) state the maximum and minimum values they would find surprising were the true number to fall outside of that range (a "nonsurprise interval"; Ranney, Munnich, and Lamprey, 2016), and (c) rate their confidence regarding their estimate. After receiving the true value, participants were also shown their original estimate and prompted to rate their surprise level on a nine-point scale. Finally, participants were queried regarding embarrassment(s) at any divergence between their estimate and the true value.

Results and discussion

Supported predictions re: Changed global warming beliefs

Ranney and Clark's (2016) central results were both replicated and extended, in terms of effect-durability. The Representative intervention replicably produced an increase in GW acceptance, while the Misleading intervention led to a decrease. Both shifts were statistically significant and numerically greater than those from Ranney and Clark's (2016): the increase was 22 percent of the possible gain (+.62: 6.79 vs. 6.17 on the nine-point scale; $t(64) = 8.069$, $p < .0001$) and larger than Ranney and Clark's 0.4-point increase (in their [representative] Experiment 6), while the *decrease* was 13 percent of the possible *loss* (−.74: 6.02 vs. 6.76; $t(59) = -6.461$, $p < .0001$)—and a larger loss compared to Ranney and Clark's 0.3-point reduction (in their [misleading] Experiment 7). Explanations for the larger upward-and-downward shifts include (a) improved methodology, (b) larger samples yielding more reliable effect sizes, and (c) a different, less homogeneous, participant population for the Misleading sample (compared to Ranney & Clark's Experiment 7 Berkeley undergraduates). The representative statistics were compelling: for *every* rare Sandwich Representative participant who decreased GW acceptance, *eight* Sandwich Representative participants *increased* their GW acceptance.

No significant decay in the heightened GW acceptance occurred between the immediate and the delayed posttest scores for both of the Representative groups, showing that the shift produced by the Representative statistics remained stable over 9 days. Indeed, the Representative statistics had still reduced the "room to improve" regarding GW beliefs by 19 percent even after nine days (also p < .0001)—barely less than the 20 percent reduction noted on the *immediate* posttest over one week earlier. (As hoped, and consistent with the effectiveness of receiving representative scientific GW information, significant *increases* in acceptance over the retention period were observed for *misleading* participants following the immediate posttest—due to the debriefing provided.)

In more complex data analyses using mixed-effects models, GW acceptance changes were even further explained with participants' (a) upon receiving the statistics' true feedback values, and (b) data from eleven demographic items (gender, age, party, religion, education, income, etc.). Space constraints prohibit elaboration, but we have noted surprise's crucial role (e.g., that greater surprise yields greater belief change; Ranney, Munnich, and Lamprey, 2016; Munnich and Ranney, 2019), and many have noted demographic variables affecting GW acceptance (e.g., McCright and Dunlap, 2011).

Partially supported predictions re: Mindset changes

Comparing pretest and immediate posttest scores, a statistically significant gain in growth mindset was found for the Sandwich Representative group, with a mean increase from 23.05 to 24.18 (out of 48; $t(64) = -3.034$, $p < .005$). However, by the delayed posttest, mindset nonsignificantly differed from pretest levels. Thus, the predicted increase in growth mindsets due to the representative statistics were only observed in the sandwich (and not the open) configuration, suggesting both (a) a metacognitive element (perhaps including experimenter demand), and (b) that the mindset change was not durable. (As noted above regarding the GW acceptance data, a more complex data analysis also showed that utilizing one's feedback-surprise rating and demographic information improved our models' explanatory power regarding mindset change.)

Once again, no polarization is observed

Polarization was absent in participants' changed GW views. Far from exhibiting a backfire effect, conservative participants receiving representative statistics showed mean GW gains at *each* conservative level (i.e., separately for those self-rating conservatism as 6, 7, 8, or 9). Likewise, liberal participants receiving the *misleading* statistics showed mean GW *losses* at *each* liberal level (i.e., separately for those self-rating as 1, 2, 3, or 4 on the 9-point conservatism scale). Indeed, both results are the same for participants for two independent self-ratings regarding economic *and* social conservatism—effectively representing an internal replication.

These findings again *refute* (1) claims that polarization occurs when people receive (at least highly germane) scientific information and (2) that such information is inert in conveying GW's reality to the public. Our interventions "floated all boats" in the directions intended, whether the information direction was representative or misleading.

Interim conclusion

Experiments 1 and 2 further demonstrated how germane, compelling, climate change information directly affects GW acceptance—in the spirit of Ranney and Clark's (2016) experiments and Mill's (e.g., 1843/2006) empiricism. In contrast, Experiment 3 breaks new empirical ground—assessing an *indirect* way to change GW attitudes.

Experiment 3: Changing GW beliefs with supra-nationalistic (and super-nationalistic) statistics

As noted above, Ranney (2012; Ranney and Thanukos, 2011, etc.) developed the RTMD theory to explain the low proportion of Americans who accept GW, relative to peer (e.g., Organisation for Economic Co-operation and Development [OECD]) nations. When the theory was conceived almost a decade ago, it was assumed that Figure 4.1's three more peripheral constructs—the acceptance of an afterlife, of high nationalism, and (their negative associate) of GW—would exhibit more modest correlations (in absolute value) among those three, compared to the remaining (12) interconstruct correlations. Early surveys suggested this to be the case, as the initial correlation between nationalism and GW acceptance across two US samples was about −.2 (Ranney, 2012).[7] Since then, though, the negative correlation between nationalism and GW acceptance seems larger in absolute value: for instance, in Experiment 1 described above, the correlation was −.43, nationalism's highest correlate between it and other RTMD constructs (and even stronger than the nationalism-deity relationship, etc., despite Figure 4.1's configuration)—which was also true in Experiment 2. GW's connection to nationalism now seems to rival the (inverse) strength of GW's connection to evolution acceptance (which was +.44 in Experiment 1), even though it was analogical aspects/relationships noticed between evolution and GW that produced RTMD theory (Ranney, 2012; Ranney and Thanukos, 2011).

Consistent with Mill's (1843/2006) interest in a contrastive study of nations, RTMD theory's central gist proposes that Americans commonly implicitly believe the United States to be the country most reinforced (for over a century) for believing that God is on its side—due to economic, military, and other successes. This high nationalism has likely been modulated by (a) often false assertions implying that the United States ranks #1 in virtually every desirable category (e.g., Congress members calling America's health care system Earth's best; cf. Davis et al., 2014) and (b) nonrandom associations among nationalism, environmental concern, and US political parties' platform-planks (e.g., the 2008 election's connection between the "Country First" slogan and the fossil-fuel-friendly "Drill, Baby, Drill" slogan at one party's political convention). More recently, the US political rhetoric even connects free-market economic positions with climate change (e.g., Lewandowsky, Gignac, and Vaughan 2013) and nationalism, such that those preferring pro-environment regulations are sometimes deemed "job-killers" and/or "anti-American."

RTMD was initially proposed based upon a set of incomplete, extant findings in the (often sociological) literature. Our laboratory's first formal surveys confirmed the directions of the predicted fifteen correlations among American participants. Experiment 3 (Luong, 2015; Teicheira, 2015) was designed to assess RTMD as a *causal* model, moving beyond correlative-explanatory accounts. This required that we manipulate a relatively benign construct to determine its effect on other constructs—particularly GW acceptance, given its societal importance. (Our laboratory already *directly* manipulated GW acceptance—as in Experiments 1 and 2, and in Ranney and Clark, 2016, etc.) Thus, this experiment manipulated nationalism level, predicting that *GW* acceptance would be indirectly changed; that is, we (partially) tested RTMD's causality by examining whether altering Americans' degrees of nationalism would drive GW acceptance changes.

It seems worth noting that our hypothesis that manipulating people's nationalism levels would change their GW acceptance is nontrivially novel, and prior to Experiment 3 (and Experiment 4), various colleagues were skeptical that the connection would prove causal—or that we could alter levels of nationalism at all. Given the past success of the NDI paradigm, though, we further hypothesized that we could bi-manipulate nationalism levels by respectively providing participants with statistics that were either pro-nationalist (e.g., on dimensions in which the United States's ranking is flattering) or *supra*-nationalist (i.e., intended to lessen super-nationalism—e.g., using dimensions in which the United States's ranking is unflattering).

Method

The experimental paradigm resembled Experiment 2's in that compelling statistics were offered to manipulate beliefs—in this case, regarding nationalistic beliefs. However, the two studies' specific methods diverged markedly in the paradigm's implementation.

Participants

Participants were US-resident MTurk workers, and 227 (35–61 per various conditions) completed the full study following exclusions.

Design, procedure, and materials

Participants were assigned to one of five groups. Groups A, B, and C received a pretest; after 12 days, they completed the study's remainder. Groups D and E received no pretest or pretreatment questionnaire, in part to assess and (if needed) account for experimenter-demand possibilities. Group A, the control group, received no treatment. Groups B and D received the pro-nationalism (+Nat) treatment, while groups C and E received the supra-nationalism (−Nat) treatment. Thus, B and C were "Sandwich" groups, and D and E were "Open" groups; Group A's mere pretest-posttest "empty sandwich" incarnation was to control for any scientific/political events occurring during the testing epoch. All groups received a posttest, and Table 4.2 illustrates the overall design.

The pretest and posttest each included a thirty-three-item survey assessing participants' attitudes regarding the six RTMD constructs. Responses used the preceding experiments' 1–9 scale. Both testings included "attention check" items, and many items, as before, were reverse-coded. Identical demographics questionnaires followed both testings.

The intervention included nine statistics chosen either to enhance pro-US nationalism (reinforcing perceptions of the United States as the world's "greatest" nation) or to reduce nationalism (employing data showing the US ranking below

Table 4.2 Experiment 3's abstracted design

		Pretest and Posttest Groups			No-Pretest Groups	
		Group A	Group B	Group C	Group D	Group E
Time 1	Test	Pretest	Pretest	Pretest		
Time 2	Statistics Provided	none (control)	+Nat	−Nat	+Nat	−Nat
	Test	Posttest	Posttest	Posttest	Posttest	Posttest

various developed nations). As in Experiment 2, participants estimated each quantity and later rated their level of surprise upon receiving each feedback value (but on a 1–5 scale, as in Munnich, Ranney, and Song, 2007). The pro-nationalist statistics favorably compared the United States to all the world's nations, whereas the supra-nationalist statistics generally compared the United States to forty-two "peer nations" (roughly aligned to the "First World")—yielding measures ranking the United States below many countries. Statistical phrasings were intended to be as apolitical as practical/possible, and the nine supra-nationalist statistics appear in Ranney, Munnich, and Lamprey (2016, p. 166, e.g., involving Americans' weight, debt, homicides, teen pregnancies, etc.).

Results and discussion

Primary analyses

The experiment successfully demonstrated that feedback on participants' statistical estimates altered their nationalism levels. Group C's supra-nationalist treatment significantly decreased nationalism by 10 percent of the possible drop (5.14 to 4.73, $t(34) = -3.3127$, $p = .0022$) and significantly increased GW acceptance by 10 percent of the possible rise (6.54 to 6.79, $t(34) = 3.441$, $p = .0015$). The pro-nationalist treatment for Group B produced a marginally significant nationalism increase (+.27 points, $p = .0649$)—and an imputed analysis involving Group D indicated a significant nationalism increase (+.39 points, $p = .0257$)—but predicted GW decreases were not statistically significant. The greater effectiveness of the supra-nationalist treatment may be attributed to the comparative familiarity of pro-nationalist arguments among many Americans (e.g., the United States having the largest GDP), as discussed more below. Two-way repeated analyses of variance (ANOVAs) confirmed, as predicted, a significant nationalism decrease and a significant GW acceptance increase in Group C, as well as a relative nationalism decrease in Group B (i.e., a significant Group × Test-time interaction was observed for nationalism: $F(1,69) = 13.04$; $p = .006$).[8]

As in prior studies, polarization was *not* observed. Conservative Group C participants—who received supra-nationalist statistics—increased their GW acceptance, as did C's liberal participants. Thus, the intervention boosting liberals' climate change beliefs did *not* have the opposite effect for conservatives.

Matching prior studies, the interconstruct correlations supported RTMD's hypothesized associations. *All* thirty correlations (fifteen pretest and fifteen posttest) were statistically significant in the directions RTMD predicts. The highest correlations (all $p < .00005$), in absolute-value terms, were Creation-Evolution (pretest: $-.875$; posttest: $-.849$), Deity-Creation (pretest: $.861$; posttest: $.879$), and Afterlife-Deity (pretest: $.888$; posttest: $.839$). Two correlations weakened significantly from pretest to posttest, in absolute terms: GW-Creation ($-.603$ to $-.445$, $p = .024$)—which may encourage communicators wishing to decouple climate change from religion or even highlight the latter's stewardship message—and Deity-Afterlife ($.888$ to $.839$, $p = .045$). Confirming similar findings that motivated this experiment, the nationalism–GW correlation was robustly about $-.35$ for both testings.

Secondary analyses

Posttest (Groups A–E) data further analyzed with fixed-effects models demonstrated that gender, age, political party, religion, and conservatism variables, when added to condition-type (i.e., pro-nationalist-, or supra-nationalist-, or no-intervention) as predictors, explained the greatest variance-percentage in both nationalism (22.2 percent) and GW acceptance (36.1 percent). The models indicated that women tended to be significantly less nationalistic than men, Libertarians were significantly less nationalistic than Democrats, and Independents were generally intermediate between Democrats and Libertarians in nationalism; all else being equal, the models indicated that Libertarians and Conservatives accepted GW less strongly than, respectively, Democrats and Liberals.

Groups receiving both a pretest and posttest (Groups A, B, and C) were yet further analyzed with mixed-effects models to assess the influence of Test (testing time) on nationalism and GW attitudes. The model employing Test, Group, the Test*Group interaction, and the demographics best explained nationalism and GW acceptance, with the *Test* predictor (a random effect in the model) representing time between pretest and posttest. Among the models including participants' *surprise ratings* (thus excluding no-treatment control Group A participants, who could not experience surprise), the full model with three two-way interactions (namely, *Demographics + Test + Group + Surprise + Test*Group + Test*Surprise + Group*Surprise*) fit best for nationalism and GW acceptance.

Significant pretest-to-posttest belief changes among RTMD constructs other than nationalism and GW could result from nationalism/GW changes impacting participants' sense of personal control—and the acceptance of awe-related experiences relating to afterlife and creation. One's control-sense could decline as one increasingly accepts GW (e.g., ruminating about how GW will affect humanity, one's self, or one's family)—producing fear, and threat/danger feelings

(Keltner and Haidt, 2003). Rutjens, van der Pligt, and Harreveld (2010) found that in losing a sense of personal control, one's acceptance of God-related roles increases to restore conceptual order. Therefore, boosting GW acceptance (by reducing nationalism) could decrease Group C's control-perception, increasing religious beliefs. However, Group B's increased nationalism (albeit a smaller effect) could raise participants' sense of control, perhaps inhibiting religious beliefs (given less need to cling to awe-related experiences), and enhancing proximal religion-conflicting scientific notions like evolution. Supporting evidence for this perspective is that (a) Group C's supra-nationalist treatment increased Afterlife acceptance (4.9→5.3, $t(34) = -2.682$, $p = .01$), whereas (b) Group B's pro-nationalist treatment increased Evolution acceptance (6.65→6.98, $t(35) = -3.004$, $p = .005$) and decreased Creation acceptance (4.46→4.09, $t(35) = 2.439$, $p = .02$). These results suggest that increasing GW acceptance with a more balanced view of a nation's strengths/shortcomings may increase a population's acceptance of applied-metaphysics beliefs involving religion, whereas focusing on national strengths may reduce creation beliefs' attractiveness. Accordingly, rather than heightened GW acceptance shaking a society's religious foundations, it might *enhance* them.

Interim conclusions

The results of the pro-nationalism and supra-nationalism interventions showed that estimated statistics followed by true-value feedback successfully changed nationalism levels. The supra-nationalist intervention's significant effects—reducing nationalism and thus boosting GW acceptance—impacted participants' nationalism more than did the pro-nationalist intervention, which did not reduce GW acceptance. This asymmetry may arise from Americans commonly hearing about US excellence, perhaps desensitizing many to additional pro-nationalistic evidence. Americans less frequently hear unflattering rankings versus peer nations, potentially explaining the supra-nationalist intervention's (e.g., Group C's) apparently greater impacts.

All thirty correlations among the six constructs over two testings were significant in the RTMD-predicted directions: GW and Evolution acceptance positively correlated, yet each negatively correlated with Nationalism, Creation, Deity and Afterlife acceptance. (The latter four constructs were positively correlated among themselves.) Again, the fifteen RTMD-predicted correlations persisted even after RTMD's nationalism construct was manipulated.

Experiment 3's considerations regarding the character of US society, among the constructs examined, seem generally coherent with the national belief-change considerations Mill (1843/2006) pondered. Decreasing one's nationalism attitudes (as with Condition C) represents a sixth way our laboratory has found to increase GW acceptance. To begin a new program of comparing the interventional efficacy among such interventions (in a *Consumer Reports* spirit), Experiment 4 sought to contrast three such ways—often in varying "dosages."

Experiment 4: Increasing GW acceptance with scientific statistics, texts, and videos

This larger, ambitious experiment simultaneously compared the effectiveness of dosages and types of brief interventions—and relevant control conditions. It also further demonstrated that quick climate instruction of various types durably increases Americans' acceptance that GW is occurring and that humankind contributes to GW. Participants received one of eleven main conditions across four types: (a) three text and (b) five video interventions that explain GW's mechanism[9] to varying degrees, (c) a replication of Experiment 2's representative-statistics' effects, and (d) two control conditions. Experiment 3's findings inspired a secondary hypothesis that increasing GW acceptance *causes* a nationalism *decrease*.

Analyses of pretests, immediate posttests, and delayed posttests roughly 9 days later showed that overall GW knowledge and acceptance had increased, with the longer-duration interventions generally resulting in larger and more durable gains. As noted below, even on the delayed posttest, every condition except the nonintervention control group showed statistically significant increases in mechanistic knowledge (cf. Thacker and Sinatra, 2019)—including the statistics condition and a unique control condition containing a "vacuous" video bereft of mechanistic explanation. Among the mechanistic interventions, those leading to greater knowledge increases generally led to greater acceptance increases. The results *disconfirm* the "science-impervious" stasis theory implicitly proposed by Kahan et al. (2012)—even beyond the disconfirmations demonstrated in Experiments 1–3 above, in Ranney and Clark's (2016) experiments, and in Ranney, Munnich, and Lamprey (2016). Further, analyses of demographic and other data yet again revealed no evidence of polarization.

Method

Experiment 4's paradigm generally resembled Experiment 2's: information was presented to change GW beliefs. The two studies' particular methods differed markedly, however, in implementation.

Participants

After 24 percent of 1447 original MTurk workers were excluded due to attention- and coherence-checks, time cut-offs, audiovisual issues, noncompletion, or several other criteria, 1103 participants' data were analyzed. Roughly half the participants' party affiliations were Republican, Libertarian, Independent, or Other—the other half representing Democratic, Green, or None.

Design, procedure, and materials

Ten interventions (including a control video)—and an eleventh, no-intervention control condition—were designed. Eight interventions included GW mechanistic instruction (varying in length and modality), one included Experiment 2's representative statistical GW information, and one was a vacuous control video offering neither mechanism nor statistics. Three mechanistic interventions were text-based: 35, 400, and 596-word written explanations. Five mechanistic interventions were videos under 1, 2, 3, 4, and 5 minutes in duration (the latter using the 596-word text as its script). Other than the control video, the interventions are at www.HowGlobalWarmingWorks.org (Ranney and Lamprey, 2013–present).

Ranney, Clark, Reinholz, and Cohen (2012b) first published the 400-word GW mechanistic explanation and its 35-word summary. Ranney et al. (2012a) and Ranney and Clark (2016) showed that the 400 words both dramatically increased participants' knowledge (consistent with Thacker and Sinatra's [2019] results) and increased anthropogenic GW *acceptance*. The 35-word text is as follows: "Earth transforms sunlight's visible light energy into infrared light energy, which leaves Earth slowly because it is absorbed by greenhouse gases. When people produce greenhouse gases, energy leaves Earth even more slowly—raising Earth's temperature."

The 596- and 400-word explanations differ (see HowGlobalWarmingWorks.org) in several ways, including added conversational elements designed to (1) inhibit self-charitable ("knew it all along") hindsight bias, (2) evoke GW's scientific consensus, (3) assure participants that *many* cannot explain GW's mechanism, and (4) motivate participants (e.g., "how can we make informed decisions without understanding the issues we're debating?"). All videos employed (a) measured, well-enunciated, narration, (b) key visual text/labels, (c) simple diagrams/animations tightly mirroring the audio, and (d) minimal peripheral design, to reduce cognitive load and keep viewers' attention on crucial aspects. Four shorter video versions were created by *nonmonotonically* reducing the 5-minute video (e.g., the 1-minute version includes phrases that are not in the 2-minute version) to those ranging from approximately 1–4 minutes.

A sixth video (Control 1) devoid of climate-scientific information was created as the first control condition by deleting mechanistic explanations and climatic evidence from the longest video. However, it included the four (1–4) elements listed above, and several other sentences from that video. The other control condition (Control 2) provided no intervention. The nine-statistics intervention was essentially the same as in Experiment 2's Representative conditions, effectively serving as a replication for that experiment—and for Ranney and Clark's (2016) Experiment 6.

Participants completed a pretest before randomly receiving one of the nine experimental interventions (text, video, statistics), Control 1, or Control 2. Half the participants (other than those in Control 2) completed an immediate posttest, yielding twenty-one total conditions. Most (68.9 percent) participants completed a delayed posttest 6–9 days later.

Each testing included GW items (generally randomized in order), including those assessing knowledge of, acceptance of, and concern about GW—as well as attitudes regarding RTMD constructs. The immediate and delayed posttests further included demographic items and items about the intervention.

During testings, participants provided written explanations regarding three aspects of GW's mechanism. The three responses per test were coded and scored together as a "response set" by two condition-blind and testing-time-blind humans using coding and scoring rubrics (Ranney and Clark, 2016, provides these, and intercoder reliabilities were comparable). A given participant's tests were all scored by the same coders, so apparent knowledge changes could not be attributed to intercoder differences.

Results and discussion

The results clearly supported a primary hypothesis—that learning scientific GW information would durably increase participants' acceptance (i.e., decrease their denial) of GW and humankind's contribution to GW. Participants exposed to short mechanistic/physical-chemical climate change instruction, or the nonmechanistic nine highly germane statistical facts, generally showed a GW acceptance increase; those participants receiving one of our nine noncontrol interventions exhibited a mean acceptance gain of 10.57 percent of the room to increase on the immediate posttest ($t(486) = 7.78$, $p < 2.1 \times 10^{-15}$), and a similarly robust 7.47 percent gain even nine days later on the delayed posttest ($t(684) = 6.73$, $p < 2.5 \times 10^{-11}$). Even the nine-statistics (i.e., "nonmechanistic") intervention, on its own, replicated Ranney and Clark (2016's Experiment 6) *and* Experiment 2's Representative effects—yielding 15.06 percent and 10.97 percent denial reductions on the immediate and delayed posttests (respectively: $t(60) = 3.0, p < .005; t(89) = 3.35, p < .001$).

Longer interventions tended to yield larger acceptance increases. The highest GW acceptance gains generally resulted from one of the three longest interventions: the 5-minute video, the 596-word text, or the nine statistics. The largest gains from pretest to *immediate* posttest resulted from the 5-minute video (20.7 percent: $t(49) = 4.58, p < 1.6 \times 10^{-5}$) and the nine statistics (15.1 percent: $t(60) = 3.0, p < .005$)—followed by the 596-word text (+11.2 percent, p < .01, which virtually tied with the 4-minute [+11.5 percent, p < .005] video). The three longest-per-mode interventions accounted for the three highest gains between pretest and *delayed* posttest, and all remained statistically significant. The 596-word text's gain was numerically the greatest after the 9 days (a robust +13.6 percent, p = .001)—its high durability perhaps reflecting the greater vividness of individuals' own mental imagery. The shortest intervention (or lowest "dose"), the 35-word text, unsurprisingly resulted in some of the smallest gains, and its result was not statistically significant on the delayed posttest, although even the 35 words yielded a marginally significant gain on the immediate posttest. In sum, we observed remarkably robust effects: For each intervention that statistically significantly decreased denial upon immediate posttesting, not one had statistically significantly *less* denial-reduction upon *delayed* posttesting. Further, the 35 words was the *only* of the nine interventions that did not obtain at least a marginally significant gain in GW acceptance after 9 days (including our original 400 words [p = .025], replicating many prior experiments' findings, e.g., Ranney and Clark, 2016).

Regarding enhancing people's GW beliefs, Experiment 4's dosage trend illustrates the importance of maintaining people's attention for longer periods (at least up to 5 minutes or 596 words; further research will determine how much more still longer interventions—e.g., Ranney and Clark's, 2016, Experiment 5—enhance effectiveness). Whether in antibiotics or cognition, more complete interventions usually outperform alternatives. While the 35-word text is better than no intervention (and useful when rehearsed and/or reflected upon), it is suboptimal for delivering practically significant GW instruction—unless its brevity facilitates providing it to a larger segment of humanity or with repetitiveness (e.g., through many 10-second Super Bowl or World Cup commercials). (Ranney, in Ranney, Munnich, and Lamprey, 2016, p. 139, wrote a thirteen-word haiku/sentence that highly concentrates GW's mechanistic description.) However, the 596-word text, the 5-minute video (Ranney et al., 2013), and the nine statistics are superior and worth the investments when possible.

The mechanistic interventions also led to large and statistically significant knowledge increases—replicating all prior experiments that employed that scientific explanation (e.g., Ranney and Clark, 2016's Experiments 2–5, and Thacker and Sinatra's recent 2019 study). The longer, more detailed explanations that tended to change GW mechanistic knowledge scores the most also tended to yield larger changes in GW acceptance. Knowledge reliably predicted acceptance at all three test times (using fixed-effects models; $p < .01$), which again (Ranney and Clark, 2016) disconfirms claims (e.g., Kahan et al., 2012) of no such relationship. The correlation between knowledge-score gains and acceptance gains was hardly perfect—perhaps because merely *recognizing* the scientific reasoning/evidence undergirding GW can enhance one's acceptance. That is, participants may recall *following* the GW explanation, *understanding* it, and *believing* it, without necessarily retaining (all) its details. This result seems satisfactory, rather like accepting "regression to the mean" without necessarily re-proving it for each use.

Replicating this chapter's prior experiments—and Ranney and Clark (2016)—Experiment 4 found no polarization. Indeed, conservatives increased their GW acceptance *more* than liberals; for instance, acceptance gains at the *delayed* posttest were significantly *positively* correlated with both conservatism measures (economic: $p < .03$; social: $p < .04$), and conservatives' gains averaged 70 percent larger than liberals' (across economic and social measures). Furthermore, at each conservatism self-rating level (from 1 to 9, including "9," extreme conservatism), participants evidenced acceptance gains. We believe such results stem from our interventions avoiding (sometimes counterbalancing) polemics or "quasi-propaganda" (even if/when the information is true) that are not rare in social psychology "vignettes"—stimuli that could possibly spawn polarization, such as language evoking inferences that experimenters may have political agendas. We focus on science and statistics, but others' interventions may offer hints of bias.

The aforementioned secondary hypothesis was also obtained, showing that nationalism and GW acceptance are *bidirectionally* causal (building on Experiment 3's result that decreasing super-nationalism increases GW acceptance). Combining all of Experiment 4's noncontrol conditions, nationalism significantly decreased ($-.09$ points; $p < .006$) on the immediate posttest, with the nine-statistics condition being a large contributor, even in isolation ($-.26$, $p < .04$). This GW-*supra*nationalism effect was largely temporally labile; few conditions showed significant nationalism decreases upon delayed posttesting, with the curious exception of the 35-word condition: –.26, p = .004. (A related, and predicted, "learning reinforcement" finding is that intervention conditions utilizing an immediate posttest yielded roughly double the GW increase on the *delayed* posttest than those without an immediate posttest [.24 vs. .12; p < .001]—so a *seemingly* diluted effect on nationalism-reduction 9 days later was expected.)

Our findings yet again disconfirm the science-impervious stasis theory that some ascribe to large swathes of Americans (implicitly or not, e.g., Kahan et al., 2012), which asserts that scientific knowledge interventions cannot increase GW acceptance, and that climate science instruction is futile or potentially counterproductive. As RTMD theory explicates, though, there are obviously potent factors (e.g., economic considerations, religious narratives, and political agendas) inhibiting GW acceptance that offer *some* resistance to scientific explanations/information (Ranney, 2012; Ranney and Thanukos, 2011). As aforementioned, such influences upon national (here, American) character is part of the research program Mill (1843/2006) seemed to call for. Fortunately for Mill's sense of progress, in Experiment 4, the receipt of scientific information was, yet again, neither insignificant nor polarizing. The findings of Experiment 4, Ranney et al. (2012a), and Ranney and Clark (2016) show that scientific interventions describing the mechanism of—or (as also in Experiment 2) representative statistical evidence regarding—GW can significantly increase climate change acceptance. These findings further *disconfirm* the stasis/science-impervious theory, highlighting the value of understanding effective interventions' characteristics.

General discussion

This chapter's four experiments, combined with Ranney and Clark's (2016) experiments, represent ten incarnations from our laboratory alone that disconfirm the stasis view that "people won't change when provided scientific information." (We have at least three new experiments that disconfirm similarly—yielding 13 or more in total, without even counting recent studies with foreign collaborators, e.g., Arnold et al., 2014.) Experiments 1–4 add to a growing set of studies demonstrating a relationship between representative scientific knowledge and GW acceptance (e.g., Shi et al., 2016)—even for the half of Experiment 2's GW stimuli that were *misleading*-yet-veridical—and even though Experiment 3's "scientific" intervention regarded America's *international status*. Furthermore, as found in our past experiments, Experiments 1–4 achieved these significant GW-denial reductions without polarizing (even economic) conservatives from liberals. Experiment 4 even demonstrated that Experiment 3's (supra-)nationalism→GW causality is *bi*directional, with increased GW acceptance (perhaps more modestly) causing decreased nationalism (i.e., exhibiting the GW→supra-nationalism direction too).

Recently—and cohering with our experiments' data (e.g., Experiment 4; Ranney, Munnich, and Lamprey, 2016's data, etc.)—Urban and Havranek (2017) independently confirmed that our 5-minute GW-mechanism video increases objective (and subjective) mechanistic knowledge. (Finding a preinstruction mechanistic GW knowledge dearth, as we always do, they also found miscalibrated impressions of individuals' understandings of GW's mechanism: those in the highest percentiles were least likely to realize that they understood it more than their peers.)

Regarding empiricists like J. S. Mill, our collective results further support a fundamental empirical notion that, even for a relatively divisive topic, merely providing short types of informative interventions (from a growing handful of such interventions) independently increases GW acceptance.[10] When one includes our laboratory's most recent finding (which space constraints prohibit us from exhibiting here) that information regarding sea-level rise also increases GW acceptance (Velautham, Ranney, and Brow, 2019), we have now demonstrated six such productive brief (e.g., roughly 5 minutes or less) information types: (1) contrastive empirical graphs, (2) empirical statistics about GW's effects (and consensus), (3) supra-nationalist empirical statistics, (4) sea-level rise information, (5) mechanistic texts, and (6) mechanistic videos. (Most of these interventions, in full or part, are available at our website, HowGlobalWarmingWorks.org; e.g., Ranney, Chang, and Lamprey, 2016.) Our laboratory is currently piloting more such interventions, including (a) why the public can/should generally trust climate scientists, (b) the flimsiness of claims that climate change is a hoax, and (c) how beneficial, both economically and in terms of health, it is for humans to switch to sustainable fuels. We note that our indirect intervention—employing supra-nationalist statistics—seems more modestly effective in increasing GW acceptance than our more direct interventions. (Similarly, we found that boosting participants' GW acceptance relatively modestly decreased their super-nationalism.)

Many ask, "Why do some deny climate change?" We answer by proposing a metaphor—a "table of denial" supported by roughly a dozen reasons or "legs."

Someone mostly denying GW may do so for only a few such reasons. They include ignorance of GW's mechanism[11], effects, or scientific consensus—which our research program addresses with our texts, videos, representative statistics (e.g., Experiments 1, 2, and 4), and (more recently and somewhat indirectly) information on sea-level rise (Velautham, Ranney, and Brow, 2019). Another "table-leg" is represented by super-nationalist thinking—including that Americans can "solve climate if need be"—which we (least directly) address with our supra-nationalist statistics (Experiment 3). Yet another denial-leg seems to be the libel that scientists are untrustworthy or hoaxers, which our current research is addressing (also see Edx.org, 2015). Likewise, our newest intervention counters a "leg" that asserts that adopting sustainable fuels is financially or societally problematic. Still other denial reasons involve scientific climate misconceptions that we plan to address. Overall, our "table-destabilizing goal" is to generate short, compelling, interventions that can "knock out" each leg of individuals' climate change denial, such that no one's "GW skepticism remains standing" (assuming that science evidence continues supporting GW acceptance!). HowGlobalWarmingWorks.org, our public-outreach website with its hundreds of thousands of pageviews (Ranney, Munnich, and Lamprey, 2016), represents a central tool in implementing that goal.

Mill's (1843/2006) interest in political ethology—particularly, a science of national character—is one many resonate to. Although few are as learned as Mill was, people are prone to generalize (e.g., "the French are X; the Chinese prefer Y; Brazilians dislike Z; Canadians are polite"). Such stereotypes often overstate national differences, sometimes relying on imprecise heuristics (e.g., availability) or unrepresentative subsamples (e.g., urban vs. rural Chinese/Bangladeshis/Brazilians). Yet Americans *do* diverge from peer nations' residents (e.g., OECD members) on various dimensions (Ranney, 2012), with modest GW acceptance being notably salient and dangerous among anomalous belief-distributions (Ranney and Thanukos, 2011). Regarding climate, humanity cannot afford delays like those preceding the acceptance of heliocentrism, a (largely) spherical Earth, or tobacco-illness connections. The exceptionalist/divergent national character exhibited by a large segment of the US population/legislatures may push Earth toward great extinctions and climate change before wisdom can expediently vanquish ignorance (Ranney, Munnich, and Lamprey, 2016).

Time is fleeting—as Anthropocene extinctions increase. We fervently hope that humans will fully comprehend anthropogenic GW extremely rapidly and collectively act to quickly inhibit climate change as optimally as possible.

The experiments above further demonstrate that accurate, representative[12], scientific information can increase the portion of the populace who accept GW's existence, magnitude, and/or threat-level—without polarization. HowGlobalWarmingWorks.org, our direct-to-the-public site for providing such information, is increasingly "international"—with burgeoning translations[13] of its texts, videos, and so on (e.g., in Mandarin, German, Czech, Spanish, Japanese, among several others); given that Earth's slender gaseous envelope is a transnational resource, we hope nations will communally redouble their efforts to address GW's potentially existential threat. Mill's century was essentially naïve concerning this threat, and largely presumed boundless frontier resources—including a rather rapidly self-cleaning ("winds will come") atmosphere. But Mill (1848/1965) saw (a) boundless growth as a danger to ecological systems, and (b) utility in nongrowing states of capital, wealth, and population. Although he quasi-celebrated diversity in national character (1843/2006), a twenty-first-century Mill would likely sacrifice bits of that diversity and urge countries to immediately reduce international variance on the dimension of sustainability. A mortal enemy of sustainability in *this* millennium is climate change.

Acknowledgments

We especially thank Charles Chang, Tina Luong, Tommy Ng, and Justin Teicheira, as well as Oliver Arnold/Taube, Lloyd Goldwasser, Jan Urban, Elijah Millgram, Florian Kaiser, Leela Velautham, Emily Harrison, Lukas Gierth, Karen Draney, Sophia Rabe-Hesketh, Daniel Wilkenfeld, Richard Samuels, Sean Li, and all of Berkeley's Reasoning Group. This research was partially supported by a bequest from the estate of Professor Barbara White.

Notes

1 Probably not surprisingly for a man of the time, Mill's writing had racist (p. 921) elements.

2 In the present chapter, *empiricist* and *rationalist* are not intended to closely track traditional philosophical usage. Instead, we use them in accord with the meanings that they have often had in psychological research (e.g., Wertheimer, 1987), and which pertain recently to the public reception of science. Roughly put, by *empiricist processes* we mean processes of belief maintenance and revision that are highly

sensitive to the data and causal models of the sort that scientific research is apt to generate. Furthermore, when saying that people are empiricists, we mean that their beliefs about the world are substantially influenced by such processes. In contrast, when speaking of *rationalistic processes*, we have in mind processes of belief maintenance and revision that are substantially *insensitive* to the data and causal models of the sort that scientific research is apt to generate. When saying that people are (not) rationalists, we mean that their beliefs about the world are (not) largely or entirely a product of such insensitive processes. So construed, it should be clear that the empiricist-rationalist distinction is both imprecise and admits of degree. Nevertheless, it is, as we will see, a useful way to organize recent debate regarding the popular reception of science, especially at it pertains to global warming (e.g., Ranney and Clark, 2016, and van der Linden et al., 2017 vs. Kahan et al., 2012). In essence, the distinction relates to the popular but oft-disconfirmed hypothesis, "People just believe what they want to believe." (One can argue that researchers who repeat this easily disconfirmed hypothesis—e.g., in suggesting that culture completely drives climate change beliefs—indirectly give comfort to those seeking to delay global warming mitigation due to personal financial interest.)

3 We thank the editors and Lije Millgram for pointing this out.
4 In philosophy, near-neighbor concepts to this extreme view might include the loyalty one finds in political noncognitivism (Millgram, 2005), religious credence (Van Leeuwen, 2014), and echo chambers (Nguyen, 2018). As the first author has pointed out elsewhere (e.g., Ranney, 2012), "facts on the ground" will likely overwhelm those denying global warming today, just as evidence such as circumnavigation and so on overwhelmed deniers of heliocentrism and/or a spherical earth in the last millennium.
5 It is clearly inaccurate both worldwide (e.g., the Paris accord) and over the long term (e.g., 1960s United States vs. today).
6 Our 1–9 scale coheres with Mill's anticipation of such measures: "beliefs . . . and the degree of assurance with which those beliefs are held" (1843/2006, p. 912).
7 A Canadian pilot study in Ontario even showed a nonsignificant positive correlation of +.08 between nationalism and global warming acceptance—probably due to a contrast effect with how the United States was viewed. This nonnegative correlation that may have already changed due to Canadian tar sand exploitation. Furthermore, Ontario differs considerably from, say, the more conservative and energy-producing Alberta, and *every* US study our laboratory had conducted demonstrates a negative nationalism-GW correlation.
8 Group A, receiving no intervention, showed no significant change over the twelve days between pretest and posttest—indicating that no external/news events altered participants' GW or nationalism views during that period.
9 Ranney and Clark (2016) explicate mechanistic information's utility in "breaking ties" between competing positions. For example, imagine asserting toilets'

existence to a skeptic in a culture that never heard of them. Drawing a bowl, a tank, pipes, and so on, and using metaphors of lakes and creeks, might "break the tie," compelling the skeptic.

10 People are clearly not fully/blindly credulous; another experiment in our laboratory, spearheaded by Leela Velautham, shows that participants were able to distinguish between the misleading and representative global warming statistics (i.e., able to identify and discriminate based on accuracy indicators), and were influenced by the informative statistics in that their global warming acceptance increased.

11 As an example of *mechanistic*, leg-destabilizing, persuasion, the first author recently asked a man who believed that increased volcanic activity was causing global warming, "Why is there increased volcanic activity at this point in earth's history?" His vacuous reply, compounded with the receipt of a science-normative (emissions) explanation, caused the man to doubt his volcanic explanation.

12 In yellow journalism's recent US upsurge, reporters and media consumers alike need greater guidance on how to detect "fake news" by assessing what is representative and what is misleading (e.g., Yarnall and Ranney, 2017).

13 Climate change acceptance is hardly unanimous outside of the United States, and even "extreme acceptors" can use help in understanding GW better to more effectively persuade others.

References

Arnold, O., Teschke, M., Walther, J., Lenz, H., Ranney, M. A., and Kaiser, F. G. (2014). Relationships among environmental attitudes and global warming knowledge, learning, and interventions. Unpublished data.

Chang, C. (2015). Bex and the magic of averaging regarding global warming. Masters project, Graduate School of Education, University of California, Berkeley.

Clark, D., Ranney, M. A., and Felipe, J. (2013). Knowledge helps: Mechanistic information and numeric evidence as cognitive levers to overcome stasis and build public consensus on climate change. In M. Knauff, M. Pauen, N. Sebanz, and I. Wachsmuth (Eds.), Cooperative Minds: Social Interaction and Group Dynamics; *Proceedings of the 35th Annual Meeting of the Cognitive Science Society* (pp. 2070–75). Austin: Cognitive Science Society.

Davis, K., Stremikis, K., Squires, D., and Schoen, C. (2014, June). Mirror, mirror, on the wall: How the performance of the U.S. health care system compares internationally. New York: The Commonwealth Fund. Retrieved from: http://www.commonwealthfund.org/~/media/files/publications/fund-report/2014/jun/1755_davis_mirror_mirror_2014.pdf

Dweck, C. S. (2006). *Mindset: The New Psychology of Success*. New York: Random House.

Edx.org/understanding-climate-denial (2015, June). *UQx Denial 101x 6.7.4.1 Full interview with Michael Ranney.* Retrieved from https://youtu.be/ElRSUgRo4dU

Garcia, J. and Koelling, R. A., (1966). Relation of cue to consequence in avoidance learning. *Psychonomic Science,* 4: 123–24.

Garcia de Osuna, J., Ranney, M., and Nelson, J. (2004). Qualitative and quantitative effects of surprise: (Mis)estimates, rationales, and feedback-induced preference changes while considering abortion. In K. Forbus, D. Gentner, and T. Regier (Eds.), *Proceedings of the Twenty-Sixth Annual Conference of the Cognitive Science Society* (pp. 422–27). Mahwah: Erlbaum.

Kahan, D. M. and Carpenter, K. (2017). Out of the lab and into the field. *Nature Climate Change,* 7: 309–11.

Kahan, D. M., Jenkins-Smith, H., Tarantola, T., Silva, C. L., and Braman, D. (2015). Geoengineering and climate change polarization: Testing a two-channel model of science communication. *Annals of American Academy of Political & Social Science,* 658: 192–222. Doi:10.1177/0002716214559002

Kahan, D. M., Peters, E., Wittlin, M., Slovic, P., Ouellette, L. L., Braman, D., and Mandel, G. (2012). The polarizing impact of science literacy and numeracy on perceived climate change risks. *Nature Climate Change,* 2: 732–35.

Keltner, D. and Haidt, J. (2003). Approaching awe, a moral, spiritual, and aesthetic emotion. *Cognition and Emotion,* 17: 297–314.

Lewandowsky, S. (2011). Popular consensus: Climate change is set to continue. *Psychological Science,* 22: 460–63.

Lewandowsky, S., Gignac, G., and Vaughan, S. (2013). The pivotal role of perceived scientific consensus in acceptance of science. *Nature Climate Change,* 3: 399–404. Doi:10.1038/nclimate1720

Lombardi, D., Sinatra, G. M., and Nussbaum, E. M. (2013). Plausibility reappraisals and shifts in middle school students' climate change conceptions. *Learning and Instruction,* 27: 50–62. Doi:10.1016/j.learninstruc.2013.03.001

Luong, T. (2015). Changing Americans' global warming acceptance with supra-nationalist statistics. Masters project, Graduate School of Education, University of California, Berkeley.

Markie, P. (2017). Rationalism vs. empiricism [July 6 revision]. *Stanford Encyclopedia of Philosophy.* Center for the Study of Language and Information (CSLI), Stanford University.

McCright, A. M. and Dunlap, R. E. (2011). Cool dudes: The denial of climate change among conservative white males in the United States. *Global Environmental Change,* 21: 1163–72.

Mill, J. S. (1843/2006). *The Collected Works of John Stuart Mill* [A System of Logic, Ratiocinative, and Inductive, Book VI, Chapter X: "Of the inverse deductive, or historical method"]. Indianapolis: Liberty Fund [reprinted of 1974's University of Toronto edition] (pp. 911–30).

Mill, J. S. (1848/1965). *The Collected Works of John Stuart Mill (Vol. III)*. [The Principles of Political Economy, Book IV, Chapter VI: "Of the stationary state."] Toronto: University of Toronto Press. http://oll.libertyfund.org/titles/243#Mill_0223-03

Millgram, E. (2005) *Ethics Done Right: Practical Reasoning as a Foundation for Moral Theory*. Cambridge: Cambridge University Press.

Munnich, E. L. and Ranney, M. A. (2019). Learning from surprise: Harnessing a metacognitive surprise signal to build well-adapted belief networks. *Topics in Cognitive Science*, 11: 164–177.

Munnich, E. L., Ranney, M. A., and Song, M. (2007). Surprise, surprise: The role of surprising numerical feedback in belief change. In D. S. McNamara and G. Trafton (Eds.), *Proceedings of the Twenty-ninth Annual Conference of the Cognitive Science Society* (pp. 503–08). Mahwah: Erlbaum.

Ng, T. K. W. (2015). The relationship between global warming and (fixed vs. growth) mindset regarding numerical reasoning and estimation. Masters project, Graduate School of Education, University of California, Berkeley.

Nguyen, C. T. (2018). Escape the echo chamber. https://aeon.co/essays/why-its-as-hard-to-escape-an-echo-chamber-as-it-is-to-flee-a-cult. Accessed on September 14, 2018.

Otto, S., and Kaiser, F. G. (2014). Ecological behavior across the lifespan: Why environmentalism increases as people grow older. *Journal of Environmental Psychology*, 40: 331–38.

Ranney, M. (1987). The role of structural context in perception: Syntax in the recognition of algebraic expressions. *Memory & Cognition*, 15: 29–41.

Ranney, M. (2008). Studies in historical replication in psychology VII: The relative utility of "ancestor analysis" from scientific and educational vantages. *Science & Education*, 17(5): 547–58.

Ranney, M. A. (2012). Why don't Americans accept evolution as much as people in peer nations do? A theory (Reinforced Theistic Manifest Destiny) and some pertinent evidence. In K. S. Rosengren, S. Brem, E. Evans, and G. M. Sinatra (Eds.) *Evolution Challenges: Integrating Research and Practice in Teaching and Learning about Evolution* (pp. 233–69). Oxford: Oxford University Press.

Ranney, M. A., Chang, C., and Lamprey, L. N. (2016). Test yourself: Can you tell the difference between a graph of Earth's surface temperature and a graph of the Dow Jones (adjusted for inflation)? [Web-pages]. In M. A. Ranney et al. (Eds.), *How Global Warming Works*. Retrieved from http://www.howglobalwarmingworks.org/2graphsab.html

Ranney, M., Cheng, F., Nelson, J., and Garcia de Osuna, J. (2001, November). Numerically driven inferencing: A new paradigm for examining judgments, decisions, and policies involving base rates. Paper presented at the Annual Meeting of the Society for Judgment & Decision Making, Orlando, FL.

Ranney, M. A. and Clark, D. (2016). Climate change conceptual change: Scientific information can transform attitudes. *Topics in Cognitive Science*, 8: 49–75. Doi: 10.1111/tops.12187

Ranney, M. A., Clark, D., Reinholz, D., and Cohen, S. (2012a). Changing global warming beliefs with scientific information: Knowledge, attitudes, and RTMD (Reinforced Theistic Manifest Destiny theory). In N. Miyake, D. Peebles, and R. P. Cooper (Eds.), *Proceedings of the 34th Annual Meeting of the Cognitive Science Society* (pp. 2228–33). Austin: Cognitive Science Society.

Ranney, M. A., Clark, D., Reinholz, D., and Cohen, S. (2012b). Improving Americans' modest global warming knowledge in the light of RTMD (Reinforced Theistic Manifest Destiny) theory. In J. van Aalst, K. Thompson, M. M. Jacobson, and P. Reimann (Eds.), *The Future of Learning: Proceedings of the Tenth International Conference of the Learning Sciences* (pp. 2-481–2-482). International Society of the Learning Sciences, Inc.

Ranney, M. A. and Lamprey, L. N. (Eds.) (2013–present). *How Global Warming Works* [Website]. Available at http://www.HowGlobalWarmingWorks.org. Accessed December 13, 2013.

Ranney, M. A., Lamprey, L. N., Reinholz, D., Le, K., Ranney, R. M., and Goldwasser, L. (2013). How global warming works: Climate change's mechanism explained (in under five minutes) [Video file]. In M. A. Ranney and L. N. Lamprey (Eds.), *How Global Warming Works*. Available at http://www.HowGlobalWarmingWorks.org/in-under-5-minutes.html. Accessed December 13, 2013.

Ranney, M. A., Munnich, E. L., and Lamprey, L. N. (2016). Increased wisdom from the ashes of ignorance and surprise: Numerically-driven inferencing, global warming, and other exemplar realms. In B. H. Ross (Ed.), *The Psychology of Learning and Motivation* (Vol. 65, pp. 129–82). New York: Elsevier. Doi:10.1016/bs.plm.2016.03.005

Ranney, M. A., Rinne, L. F., Yarnall, L., Munnich, E., Miratrix, L., and Schank, P. (2008). Designing and assessing numeracy training for journalists: Toward improving quantitative reasoning among media consumers. In P. A. Kirschner, F. Prins, V. Jonker, and G. Kanselaar (Eds.), *International Perspectives in the Learning Sciences: Proceedings of the 8th International Conference for the Learning Sciences* (pp. 2-246–2-253). International Society of the Learning Sciences, Inc.

Ranney, M. and Schank, P. (1998). Toward an integration of the social and the scientific: Observing, modeling, and promoting the explanatory coherence of reasoning. In S. Read and L. Miller (Eds.), *Connectionist Models of Social Reasoning and Social Behavior* (pp. 245–74). Mahwah: Lawrence Erlbaum.

Ranney, M., and Thagard, P. (1988). Explanatory coherence and belief revision in naive physics. *Proceedings of the Tenth Annual Conference of the Cognitive Science Society* (pp. 426–32). Hillsdale: Erlbaum.

Ranney, M. A., and Thanukos, A. (2011). Accepting evolution or creation in people, critters, plants, and classrooms: The maelstrom of American cognition about biological change. In R. S. Taylor and M. Ferrari (Eds.), *Epistemology and Science Education: Understanding the Evolution vs. Intelligent Design Controversy* (pp. 143–72). New York: Routledge.

Rutjens, B. T., van der Pligt, J., and van Harreveld, F. (2010). Deus or Darwin: Randomness and belief in theories about the origin of life. *Journal of Experimental Social Psychology*, 46: 1078–80.

Sahakian, W. S. (1968). *History of Psychology: A Source Book in Systematic Psychology*. Itasca, IL: Peacock.

Shi, J., Visschers, H. M., Siegrist, M., and Árvai, J. (2016). Knowledge as a driver of public perceptions about climate change reassessed. *Nature Climate Change*, 6: 759–62.

Teicheira, J. (2015). Increasing global warming acceptance through statistically driven interventions. Masters project, Graduate School of Education, University of California, Berkeley.

Thacker, I. and Sinatra, G. M. (2019). Visualizing the greenhouse effect: Restructuring mental models of climate change through a guided online simulation. *Education Sciences*, 9(1), 14. Retrieved from http://dx.doi.org/10.3390/educsci9010014

Thagard, P. (1989). Explanatory coherence. *Behavioral and Brain Sciences*, 12: 435–502.

Urban, J. and Havranek, M. (2017, September). Can mechanistic explanation make people with low level of knowledge see their ignorance? Paper presented at the European Society for Cognitive Psychology, Potsdam.

van der Linden, S., Maibach, E., Cook, J., Leiserowitz, A., Ranney, M., Lewandowsky, S., Árvai, J., and Weber, E. (2017). Culture vs. cognition is a false dilemma. *Nature Climate Change*, 7: 457.

Van Leeuwen, N. (2014). Religious credence is not factual belief. *Cognition*, 133: 698–715.

Velautham, L., Ranney, M. A., and Brow, Q. (2019). Communicating climate change oceanically: Sea level rise information increases mitigation, inundation, and global warming acceptance. *Frontiers in Communication*, 4, 1-17 [article 7; 17 pages]. https://doi.org/10.3389/fcomm.2019.00007

Vice, M. (2017). Publics worldwide unfavorable toward Putin, Russia. Pew Research Center. Available at http://www.pewglobal.org/2017/08/16/publics-worldwide-unfavorable-toward-putin-russia/. Accessed August 19, 2007.

Wertheimer, M. (1987). *A Brief History of Psychology*. New York: Holt, Rinhart, & Winston.

Williamson, S. (2015). Daily closing values of the DJA in the United States, 1885 to present. Retrieved October 22, 2014, from: http://www.measuringworth.com/DJA/

Yarnall, L. and Ranney, M. A. (2017). Fostering scientific and numerate practices in journalism to support rapid public learning. *Numeracy*, 10(1): 1–28 [article 3; 30 pages]. doi: http://dx.doi.org/10.5038/1936-4660.10.1.3 Also available at: http://scholarcommons.usf.edu/numeracy/vol10/iss1/art3

5

Doubly Counterintuitive: Cognitive Obstacles to the Discovery and the Learning of Scientific Ideas and Why They Often Differ

Andrew Shtulman

Collectively, humans know more about how the world works than ever before. This knowledge is the hard-won achievement of innumerable scientists across innumerable years. Their labors include designing instruments of measurement, devising experimental protocols, recording and disseminating data, constructing theoretical accounts of those data, debating the merits of different theoretical accounts, and unifying insights across disparate paradigms or fields. The goal of these activities was to create ever-more accurate and ever-more coherent models of reality—models used to inform technological innovation and edify future generations.

Despite these collective advances in human knowledge, most individual humans know very little about science. Organizations like Gallup, the Pew Research Center, and the National Science Foundation have been polling the general public on their understanding of science for decades and have documented consistently low levels of scientific literacy. For instance, a research survey by the Pew Research Center (2015) found that only 65 percent of Americans believe that humans have evolved over time, compared with 98 percent of the members of the American Association for the Advancement of Science (AAAS); only 50 percent of Americans believe that climate change is due mostly to human activity, compared with 87 percent of AAAS members; and only 37 percent of Americans believe that genetically modified foods are safe to eat, compared with 88 percent of AAAS members.

Poor science education is one reason the average person knows little science, but it is not the only reason. Decades of research on science education have revealed that individuals exposed to extensive and comprehensive science instruction often fail to learn from it (e.g., Gregg et al., 2001; Kim and Pak, 2002; Libarkin and Anderson, 2005; Shtulman and Calabi, 2013). Students enter the science classroom with naïve, nonscientific ideas about how the world works, and they leave the classroom with those same ideas intact. Instruction is ineffective because science is deeply counterintuitive. Science defies our earliest and most accessible intuitions about how the world works, and those intuitions impede our ability to acquire more accurate models of the world. Learning science is of course possible, but the process is difficult and protracted (Carey, 2009; Shtulman, 2017; Vosniadou, 1994a).

This tension, between the advancement of science as a whole and the learning of science by individuals, has implications for the study of scientific knowledge. Scholars interested in the origin and character of scientific ideas are likely to learn different lessons from the professional activities of scientists than from the cognitive activities of science students. Our generalizations about scientific knowledge will differ depending on whom we take as the custodians of that knowledge and whose struggles we view as most informative. Students' knowledge cannot be written off as a corrupted or degraded version of scientists' knowledge, because the two forms of knowledge may embody different relations among scientific concepts or different relations between scientific concepts and empirical observations.

Here, I explore a question common both to the scientist's struggle to model reality and the student's: why are some scientific ideas particularly difficult to grasp? Atoms, germs, heat, inertia, heliocentrism, natural selection, continental drift: these ideas were slow to develop in the history of science and remain slow to develop in the minds of individuals, but the reasons for the historical delay are not necessarily the same as the reasons for the cognitive delay. Scientists and students have different explanatory goals, different empirical concerns, and different background assumptions, and I aim to show how these factors can render the same idea counterintuitive for different reasons. This comparison of scientists' and students' conceptual ecologies has implications not only for theories of scientific knowledge but also for the practice of teaching science to nonscientists.

Two caveats should be noted. First, my focus is on the content of scientific claims rather than the process of testing those claims. Much of the psychological research on scientific thought examines inquiry skills, like the ability to design informative experiments (Kuhn and Pease, 2008; Lorch et al., 2010), the ability to evaluate empirical data (Chinn and Brewer, 1998; Morris and Masnick, 2014), or the ability to coordinate data and theory (Bonawitz et al., 2012; Schauble, 1996). Scientists may engage in more sophisticated inquiry activities than students, but the focus of this chapter will be on the products of those activities—the concepts and theories informed by inquiry—and how those products are understood.

Second, my focus is on what makes scientific *ideas* counterintuitive for scientists and students, rather than how those ideas are constructed. The process of constructing scientific ideas often entails conceptual change, or knowledge restructuring at the level of individual concepts (Carey, 2009; Chi, 2005; Nersessian, 1989), and there is debate as to how closely conceptual change in the student mirrors conceptual change in the scientific community (see, for example, DiSessa, 2008; Kuhn, 1989). Differences in the process of conceptual change may yield differences in the outcome of that process, but the focus here will be on the latter. That is, I will focus on what makes scientific ideas difficult to grasp rather than on how we come to grasp them.

A common starting point: Intuitive theories

Humans are built to perceive the environment in ways that enhance survival, which do not always align with the categories of science (Carey and Spelke, 1996; Chi et al., 2012). These misalignments can take the form of omissions or commissions. The omissions are when we fail to perceive the entities or processes causally responsible for some phenomenon, whereas the commissions are when we mistakenly assume that a phenomenon is caused by entities or processes we can perceive.

Errors of omission are particularly common when reasoning about biology. Biological systems usually operate at too small a scale for us to observe firsthand. We cannot observe the functional relations among internal organs or the genetic underpinnings of heritable traits, so we gravitate toward generic explanations of metabolism and inheritance, such as *vitalism*, or the belief that organisms possess an internal life-force that maintains growth and health (Inagaki and Hatano, 2004; Morris, Taplin, and Gelman, 2000); and *essentialism*, or the belief that an organism's external properties are determined by an internal essence inherited at birth (Gelman, 2003; Johnson and Solomon, 1997).

Errors of commission, on the other hand, may be more common when reasoning about physics. Physical systems are multifaceted, and we observe only the facets that impact our interaction with the system. For instance, we perceive material objects in terms of their heft (felt weight) and bulk (visible size), not their actual weight and size (Smith, Carey, and Wiser, 1985; Smith, 2007), and we perceive gravity as pulling us down, not pulling us toward the center of the Earth (Blown and Bryce, 2013; Vosniadou and Brewer, 1994).

These perceptual biases lead humans down the wrong path when it comes to theorizing about the causes of natural phenomena, pushing us to draw distinctions that are not particularly meaningful from a scientific point of view (e.g., a distinction between motion and rest) and to overlook distinctions that *are* meaningful (e.g., a distinction between weight and density). What's more, they lead everyone down the wrong path, students and scientists alike. Students' preinstructional beliefs in many domains resemble the first theories to emerge in the history of science. For example, students' preinstructional beliefs about motion resemble the "impetus theory" of the Middle Ages more closely than Newtonian mechanics (McCloskey, 1983). Their beliefs about inheritance resemble Lamarck's theory of acquired characters more closely than a genetic theory (Springer and Keil, 1989). And their beliefs about astronomy resemble pre-Copernican models of the solar system more closely than post-Copernican ones (Vosniadou and Brewer, 1994).

The beliefs of today's students resemble those of yesterday's scientists in both form and function. Physics students, for instance, make essentially the same predictions about motion that Medieval physicists made, and they provide essentially the same explanations (Eckstein and Kozhevnikov, 1997; McCloskey, 1983). Across tasks and contexts, their beliefs about motion are generally as coherent as Medieval physicists', which is one reason psychologists terms those beliefs *theories*. Another reason is that the beliefs facilitate the same cognitive activities as scientific theories: explaining past events, predicting future events, intervening on present events, and reasoning about counterfactual events (Gelman and Legare, 2011; Gopnik and Wellman, 2012; Shtulman, 2017).

Intuitive theories are the starting point for how humans represent and understand the natural world, emerging early in life and in similar forms across cultures (Shtulman, 2017). They shape the foundations of everyday reasoning, including the foundations of scientific inquiry. However, the historical pathway from intuitive theories to scientific theories is often quite different from the developmental pathway. Scientists revise their theories through iterative cycles of data collection and data interpretation, whereas nonscientists typically maintain the same intuitive theory until confronted with a scientific alternative. Students are the beneficiaries of a vast effort to vet empirical ideas without doing any of the vetting. But the vetting, I will argue, may change scientists' understanding of the role and value of the vetted product.

Divergent paths: Discovering vs. learning scientific truths

Scientific truths can be difficult to discover for different reasons than they are difficult to learn. Here, I will sketch some ways in which the motivations and assumptions of the discoverers (scientists) differ from those of the learners (students) and how those differences can shape the cognitive obstacles to embracing a scientific truth. The contrast I draw between discoverers and learners is not meant to be holistic, in the sense that some people are discoverers and others are learners, but rather concept-specific. Discoverers are those who first formulate a scientific concept, and learners are those who are introduced to the concept secondhand.

Divergent explanatory goals

When scientists construct a new theory, their primary goal is to account for an existing body of data, but other goals are pursued as well. Scientists try to maximize explanatory scope, minimize auxiliary assumptions, generate new hypotheses, avoid internal inconsistencies, and be consistent with established theories in related domains (Laudan et al., 1986; Thagard, 1978). Nonscientists care about these additional considerations when they are directly asked to compare two theories (Koslowski et al., 2008; Samarapungavan, 1992), but there is little evidence that these considerations inform the construction of intuitive theories. Intuitive theories are a response to everyday phenomena—motion, heat, weather, illness, growth—and are constructed primarily to account for how we perceive those phenomena. Higher-order considerations like generativity, parsimony, and breadth may implicitly guide the construction of intuitive theories but do not seem to be engaged explicitly (DiSessa, 1993; Kuhn, 1989).

Consider the domain of matter. The idea that objects are composed of microscopic particles—atoms—was debated within the scientific community for hundreds of years (Toulmin and Goodfield, 1962) and is not fully embraced by children until the second decade of life (Smith, 2007). Early chemists agreed that material objects were composed of more fundamental elements, but they disagreed about what those elements were, what shape they took, how they interacted, whether they could be transmuted from one type to another, and whether they were the constituents of all entities or only inorganic entities. Early chemists also disagreed about the relations between matter and space, matter and motion, matter and sensation, and matter and mind. The disagreements were moral as well as empirical. Early skeptics of atomism saw bleak implications in the claim that reality consists of nothing more than atoms and void, fearing it implied a lack of purpose and design.

The goals of early chemists thus extended beyond the domain of matter into the domains of biology, psychology, and ethics. The goals of nonchemists, on the other hand, are much simpler: to account for everyday material transformations like sinking and floating, shrinking and expanding, freezing and burning. These transformations occur at a macroscopic level, but they are constrained by processes operating at a microscopic level, and nonchemists have difficulty relating the two levels, preferring to interpret material transformations as directed processes rather than emergent phenomena (Chi et al., 2012). Thus, while early chemists were reluctant to embrace atomic theory for metaphysical reasons, nonchemists are reluctant to do simply because they cannot see microscopic particles or fathom how such particles could give rise to outcomes they can see (Chi, 2005).

Another example of divergent explanatory goals can be seen in the domain of motion. A guiding principle for sixteenth- and seventeenth-century physicists, like Kepler and Newton, was to account for terrestrial motion (e.g., an apple falling from a tree) and celestial motion (e.g., a moon orbiting a planet) with the same laws (Holton, 1988). Beginning with the Greeks, terrestrial motion and celestial motion were treated as separate phenomena and explained by separate principles. Even terrestrial motion was subdivided into separate phenomena— flinging vs. falling vs. spinning—and explained by separate principles. Physicists like Kepler and Newton sought to unify all motion with a single mechanics.

Nonphysicists, on the other hand, are concerned not with explaining motion in general but with explaining particular instances of motion: the trajectory of a ball off a bat, the speed of a sled down a hill, the arc of water in a drinking fountain (DiSessa, 1993). The nonphysicist wants to know where these things are going and how quickly. Little effort is spent comparing one instance of motion to another, leading to discrepant predictions. A ball rolled off a cliff is expected to move forward as it falls, but a ball dropped from a plane is expected to fall straight down. Both have horizontal velocity and thus both would move forward as they fell. However, nonphysicists attribute a force to the first ball—the "force of motion"—but attribute no force to second (Kaiser, Proffitt, and McCloskey, 1985). Nonphysicists are hard-pressed to see the similarity between a ball that is dropped and a ball that is rolled, whereas many early physicists saw the similarity but were hard-pressed to explain it.

Divergent empirical concerns

Just as scientists approach theory construction with a wider range of goals than nonscientists, they are also aware of a wider range of phenomena to be covered by those theories, including anomalies. Anomalies are observations that cannot be explained by a field's prevailing theory, such as the observation that Uranus does not follow a perfectly elliptical path (leading to the discovery of Neptune) or the observation that some metals gain weight when burned (leading to the discovery of oxygen). They play an important role in scientific innovation (Kuhn, 1962), but nonscientists know nothing of them. Nonscientists struggle to account for everyday observations, whereas scientists struggle to account for both everyday observations and anomalies—the latter typically discovered through careful, systematic observation.

In the domain of heat, an anomalous finding that spurred scientific innovation was Black's discovery that heat and temperature are dissociable at a phase change (Fox, 1971; Wiser and Carey, 1983). Black observed that adding heat to a mixture of ice and water did not raise its temperature but rather increased the proportion of water to ice. Only after all the ice had melted did the water's temperature rise. Black observed the same pattern for a mixture of boiling water and steam; adding heat to this mixture did not raise its temperature until all the water had turned to steam. Chemists before Black had assumed that thermometers measure heat, not temperature, and they had no explanation for how heat could be added to a physical system without a concurrent change in temperature. Black posited a new substance—caloric—to explain his findings. Caloric was believed to pool inside substances at their melting point or boiling point, changing the substance's chemical composition but not its temperature.

Caloric is a fiction; the correct explanation for why heating a substance does not increase its temperature at a phase change is that the added energy is spent breaking molecular bonds. This explanation requires thinking of heat as kinetic energy at the molecular level. Black did not think of heat this way, and neither do nonchemists today. But nonchemists are generally unaware of the thermal dynamics of phase change and are thus uncompelled to explain them (Wiser and Amin, 2001).

For nonchemists, the phenomena most in need of explanation are the physical sensations of warmth and cold. Warmth and cold are reified either as properties intrinsic to matter or as substances that flow in and out of matter (Erickson, 1979; Reiner et al., 2000). This view leads to a conflation of heat and temperature, as well as the misconception that heat and cold are distinct kinds of substances. This set of beliefs overlaps with Black's theory of caloric in some regards but not others. It leads non-chemists to construe heat as a kind of substance, as Black did, but it also leads them to conflate heat with temperature, which Black did not.

An additional example of the divergent concerns of scientists and students comes from astronomy. For centuries, the Earth was believed to be at the center of the universe, but this model of the universe could not easily account for early observations of planetary motion (Toulmin and Goodfield, 1961). Viewed from Earth, the planets appear to move backward for several weeks of their orbit. Early astronomers like Hipparchus and Ptolemy accounted for this anomaly by positing a complicated system of epicycles, or small circles traversed by each planet along their larger circle around the Earth.

Eventually, this convoluted, geocentric model was replaced with a simpler, heliocentric model, but the heliocentric model remains counterintuitive to nonastronomers. Only around 75 percent of Americans accept that the Earth revolves around the sun; the remaining 25 percent believe that the sun revolves around the Earth (National Science Board, 2014). Nonastronomers are reluctant to embrace heliocentrism not because they are committed to epicycles but because they perceive the sun as rising and setting, and they do not perceive the Earth as rotating or revolving (Harlow et al., 2011; Vosniadou and Brewer, 1994). Accounting for the sun's apparent motion is the chief concern of nonastronomers, who neither perceive nor consider the motions of other planets.

Divergent background assumptions

Scientific ideas are constrained by a host of background assumptions about how the world works in general. The assumptions that make an idea counterintuitive to a scientist, steeped in particular methodological and theoretical traditions, may be quite different from those that make the same idea counterintuitive to a nonscientist. The sticking points for scientists may not be non-issues for nonscientists and vice versa.

Consider the claim that the Earth's continents move. On first blush, nonscientists view this claim as absurd. They see the Earth as essentially an inert chunk of rock, solid and eternal (Libarkin et al., 2005; Marques and Thompson, 1997). Accepting that the continents move requires reconceptualizing the Earth itself, from a static object characterized by small, inconsequential changes (like eroding mountains and shifting coastlines) to a dynamic system characterized by large, continual change (like sinking landmasses and colliding plates).

Geologists, on the other hand, were initially resistant to this idea for different reasons. They did not view the Earth as static. They knew, by the time that Wegener proposed his theory of continental drift, that the Earth had begun its existence in a molten state, that its interior was still hotter and more fluid than its exterior, that its oceans were once vaster, and that its mountains were once flatter. What made geologists of the early twentieth century skeptical of Wegener's theory is that they were unable to reconcile the theory with its implications. They were willing to concede that the Earth's crust could crack or fold, but they were unwilling to concede that it could rearrange itself into new configurations, for they knew of no mechanism that would allow whole continents to move (Oreskes, 1999; Gould, 1992).

Tellingly, a key piece of evidence that convinced geologists that the continents move was the discovery of magnetic stripes on the seafloor. These stripes indicate that currents of molten rock deep within the Earth's interior have changed the magnetic properties of the Earth's crust, as that crust forms anew at the boundaries of tectonic plates (Oreskes, 1999). Nongeologists do not know of the existence of magnetic stripes, let alone appreciate their implications.

The domain of illness provides another example. Many, if not most, of the illnesses that plague humanity are caused by microbial infection. Microbes, or germs, cannot be perceived, nor can they be tracked in their transmission from one host to another, so their discovery took centuries. The perceptual obstacles to identifying germs were compounded by conceptual ones. How could something alive be too small to be seen? How could one living thing survive and reproduce inside another? Nonbiologists continue to be puzzled by such dilemmas. They now know about the existence of germs—even preschoolers know that germs make a person sick (Kalish, 1996) and that germs spread on contact with an infected individual (Blacker and LoBue, 2016)—but they do not conceive of illness as the biological consequence of a parasite hijacking the host's resources to further its own survival and reproduction (Au et al., 2008).

The historical discovery of germs was also hampered by considerations that nonbiologists do not entertain: that the body contains a supply of internal fluids, or "humors," whose balance is critical for health and vitality (Lederberg, 2000; Thagard, 1999). Beginning with Hippocrates, early physicians analyzed disease as the interplay between blood (the sanguine humor), phlegm (the phlegmatic humor), yellow bile (the choleric humor), and black bile (the melancholic humor). Too much blood was thought to cause headaches; too much phlegm, epilepsy; too much yellow bile, fevers; too much black bile, depression. The prescribed cures were to relieve the body of the excess humor by inducing vomiting, defecation, or bleeding.

This framework made the notion of microbial infection even more problematic. Early biologists were willing to accept that humors could become imbalanced by external factors—notably, bad air or "miasma"—but the true cause of illness was the imbalance, not the imbalancer. External factors were not construed as sources of contagion. Once again, a telling sign of the difference between scientists and nonscientists' acceptance of the correct theory comes from a discovery that only scientists would find convincing: the discovery that fermentation of wine requires a living organism—yeast—which consumes the sugar in grapes and excretes alcohol as waste. Yeast was an existence proof for nineteenth-century biologists of how a foreign microbe could alter the functioning of a biological system, but most nonbiologists remain unaware that yeast is alive (Songer and Mintzes, 1994), let alone the correspondence between yeast's role in fermentation and a pathogen's role in human illness.

Case study: Divergent paths to understanding evolution

Evolution by natural selection is a prime example of a theory that was counterintuitive to early scientists for different reasons than it is counterintuitive to science students. Here, I will outline key differences in the motivations and assumptions of early evolutionary theorists and those who learn about evolution secondhand. I focus on differences in background assumptions, as these differences are perhaps the most important factor in this domain, theoretically and pedagogically.

Divergent explanatory goals

The idea that species change over time was entertained as early as antiquity, but it was not widely investigated until the eighteenth and nineteenth centuries

(Bowler, 1992; Mayr, 1982). Biologists of that period had amassed a large database of specimens and sought a naturalistic explanation for the origin of species and their adaptation to particular environments. The traditional explanations were *divine creation*, the idea that species were created in their present form by a divine power; and *spontaneous generation*, the idea that species emerged from the Earth whole-cloth. Neither provided a generative framework for empirical inquiry.

Biologists dissatisfied with creationism and spontaneous generation revisited the idea that the Earth's species had not always existed but were instead the descendants of some smaller number of ancestral species. It was agreed that the ancestral species spread and diversified, but it was debated whether this happened because of environmental circumstances or because of some inherent property of living things. For eighteenth- and nineteenth-century biologists, evolution was an accepted possibility—and a preferred alternative to supernatural explanations—but the mechanism was a mystery.

For nonbiologists, on the other hand, evolution itself is an unlikely supposition (Blancke et al., 2012), as it is not readily observed nor inferred. Casual observation of plants and animals conveys no impression that they have changed over time. The biological world appears to be as static and eternal as the Earth itself. The pressing biological questions for nonbiologists are not where species come from and why they are adapted to their environment but which species are safe to interact with and which should be avoided (Barrett and Broesch, 2012; Wertz and Wynn, 2014). The very idea of evolution has to be suggested by others; it is not intuited as a possibility (Shtulman, Neal, and Lindquist, 2016).

Another reason evolution is viewed as irrelevant to everyday biological concerns is that nonbiologists already have an explanation for adaptation and speciation: divine creation (Heddy and Nadelson, 2012; Newport, 2010). Divine creation is endorsed by individuals of varying ages and upbringings, including children raised in secular households (Evans, 2001). When elementary schoolers are asked where the first bear came from or where the first lizard came from, they usually say that God created them, even when their own parents say that bears and lizards evolved from earlier forms of life. Divine creation embodies a form of causation we are all familiar with—intentional design—and it is thus preferred to a more complicated explanation like evolution. Evolution may provide a naturalistic account of the origins of life—a primary desiderata for biologists—but nonbiologists are generally unperturbed by the supernatural aspects of creationism. Most people consider supernatural causes to be just as plausible as natural ones (Legare and Shtulman, 2018).

Divergent empirical concerns

Eighteenth- and nineteenth-century biologists knew of a wide range of empirical phenomena that current students do not. They knew of extinct species, through their fossilized remains, and wondered how those species are related to extant species. They knew of *analogous* traits, or traits with similar functions but dissimilar structures, such as bat wings and bird wings, and wondered how those traits emerged seemingly independently. They knew of *homologous* traits, or traits that have taken on new functions or lost their old functions, such as the blind mole rat's eye or the human tailbone, and wondered whether those traits are the remnants of a shared body plan. These facts constrained early theories of evolution and even suggested possible mechanisms. For instance, widespread homologies across species motivated Cope's theory of accelerated growth, or the theory that evolution results from the acceleration and compression of universal stages of embryonic growth, with new stages added on top of old ones (Bowler, 1992).

Nonbiologists are generally unaware of these facts. They may know of fossils and shared traits, but they do not necessarily see these phenomena as evidence of evolution (Evans et al., 2010). The primary evidence of evolution for a nonbiologist is public discourse about evolution and public representations of evolution. The discourse includes claims about common ancestry (e.g., that humans and chimps share 98 percent of their DNA), claims about adaptation (e.g., that white fur is an adaptation to Arctic climates), and the controversy over teaching evolution in school. The public representations include evolutionary trees, nature documentaries, cartoons, and even video games (e.g., *Spore*, *SimEarth*, *Pokémon Go*). Nonbiologists learn about evolution not through observation but through culturally transmitted information, and the challenge for nonbiologists is interpreting this information, which is often vague or misleading.

Evolutionary trees are a prime example of misleading information. Evolutionary trees depict *speciation*, or the emergence of new species. Speciation is inherently a branching process, of one species diverging from another, but it is often depicted as a linear process, of one species giving rise to another, by the evolutionary trees in textbooks and science museums (Catley and Novick, 2008; MacDonald and Wiley, 2012). The nodes in these trees are labeled with extinct species, implying that they gave rise to the extant species along the trees' tips, which is highly unlikely given the ubiquity of extinction. Other problematic features of evolutionary trees include varying the thickness of a tree's branches without explanation, varying the endpoints of a tree's branches without explanation, segregating "higher" organisms from "lower" organisms, and placing humans on the top-most branch of a vertically arrayed tree or the right-most branch of a horizontally arrayed tree (Catley and Novick, 2008; MacDonald and Wiley, 2012; Shtulman and Checa, 2012).

Evolutionary trees, and other popular depictions of evolution, thus present significant interpretive challenges to nonbiologists. Whereas early evolutionary theorists struggled to interpret varied traces of evolution in the fossil record and the zoological record, nonbiologists struggle to interpret ambiguous or misleading representations of evolution in the public record.

Divergent background assumptions

Darwin's discovery of the principle of natural selection revolutionized the biological sciences. While Darwin was one of many biologists trying to understand speciation and adaptation from a naturalistic point of view, he was one of the first to realize that evolution proceeds via selection over a population. Darwin's predecessors and contemporaries had posited many mechanisms of their own—the inheritance of acquired characters (Lamarck's mechanism), the law of accelerated growth (Cope's mechanism), the inherent properties of organic matter (Eimer's mechanism)—but all such mechanisms operated indiscriminately, propelling evolution in each and every lineage of living things. Darwin, on the other hand, realized that evolution is an emergent property of the selective survival of only some lineages within a population.

From where did this insight arise? The history of science suggests that three events were critical: (1) Darwin's journey to the Galapagos, which opened his eyes to the ubiquity of variation within a species (Lack, 1947/1983); (2) Darwin's reading of Malthus's *Essay on the Principle of Population*, which opened his eyes to resource limitation and its role in inciting competition within a population (Millman and Smith, 1997); and (3) Darwin's reading of Lyell's *Principles of Geology*, which opened his eyes to the transformative power of incremental change over vast periods of time (Gruber, 1981).

These events instilled in Darwin an appreciation of intraspecific variation, intraspecific competition, and geologic time, respectively. All were important to Darwin's theorizing, but one concept in particular—intraspecific variation—has been implicated as his most important insight. Philosophers of biology commonly argue that what set Darwin's theory apart from his peers' was that Darwin's was population-based whereas those of his peers were typological. Darwin treated species as continuums of variation whereas his peers treated species as discrete, homogenous types (Gould, 1996; Hull, 1965; Mayr, 1982; Sober, 1994).

Students of biology today have difficulty understanding the same three concepts that proved critical to Darwin's theorizing. Students view variation between species as pervasive and adaptive but variation within species as minimal and nonadaptive (Nettle, 2010; Shtulman and Schulz, 2008). They claim, for instance, that most traits appear in duplicate form across the entire species and that it is unlikely a member of the species could be born with a different version of the trait. Students also hold overly simplistic views of the relations among organisms within an ecosystem—views that downplay competition for resources between species, let alone within species (Özkan, Tekkaya, and Geban, 2004; Zimmerman and Cuddington, 2007). Most believe that stable ecosystems are characterized by ample food, water, and shelter, and that all inhabitants of the ecosystem are able to survive and reproduce. Lastly, students underestimate the duration of geological events by several orders of magnitude (Lee et al., 2011; Trend, 2001). They date the origin of mammals hundreds of millions of years too close to present day and the origin of life billions of years too close to present day.

A psychological question motivated by the history and philosophy of biology is whether understanding evolution by natural selection requires an understanding of all three concepts—intraspecific variation, intraspecific competition, and geologic time—or whether one concept in particular is most critical, namely, intraspecific variation. In my lab, we explored this question directly, surveying students' understanding of variation, competition, and time in relation to their understanding of evolution (Shtulman, 2014). The students were recruited from introductory psychology courses, and they reported having taken an average of 1.2 college-level biology courses. Some were biology majors, but most were not.

We assessed students' understanding of evolution using a battery of questions designed specifically to differentiate population-based reasoning from typological reasoning (Shtulman, 2006). The questions covered six evolutionary phenomena—variation, inheritance, adaptation, domestication, speciation, and extinction—and solicited a combination of closed-ended and open-ended responses.

Here is a sample question regarding adaptation: "A youth basketball team scores more points per game this season than they did the previous season.

Which explanation for this change is most analogous to Darwin's explanation for the adaptation of species? (a) Each returning team member grew taller over the summer; (b) Any athlete who participates in a sport for more than one season will improve at that sport; (c) More people tried out for the same number of spots this year; (d) On average, each team member practiced harder this season." The correct answer is (c), as it is the only answer that evokes selection, but most survey respondents chose one of the other answers, which evoke mechanisms that operate on the group as a whole. And those who chose (a), (b), or (d) as most analogous to Darwin's explanation for adaptation typically chose (c) as *least* analogous, further indicating that they do not see selection as relevant to evolution.

Scores on this survey, in its entirety, could range from −30 to +30, with negative scores indicating typological reasoning and positive scores indicating population-based reasoning. In actuality, they ranged from −25 to +24, with an average score of −2.3.

To measure students' understanding of intraspecific variation, we adapted a task from Shtulman and Schulz (2008). Participants were asked whether each of three traits—a behavioral trait, an external anatomical trait, and an internal anatomical trait—could vary for each of six animals. Half the animals were mammals (giraffes, pandas, kangaroos) and half were insects (grasshoppers, ants, bees). One trial pertained to kangaroos having two stomachs. For this trial, participants were told, "It is commonly observed that kangaroos have two stomachs," and they were then asked (1) "Do you think all kangaroos have two stomachs or just most kangaroos?" and (2) "Could a kangaroo be born with a different number of stomachs?" Across species and traits, participants judged traits actually variable (question 1) 47 percent of the time and potentially variable (question 2) 61 percent of the time.

To measure participants' understanding of intraspecific competition, we presented participants with sixteen behaviors and asked them to indicate which of six animals exhibit that behavior. The behaviors came in four types: cooperation within a species (e.g., nursing the offspring of an unrelated member of the same species), cooperation between species (e.g., sharing a nest or burrow with an animal from a different species), competition within a species (e.g., eating another member of the species), and competition between species (e.g., tricking an animal from a different species into raising one's young). We paired the properties with unfamiliar animals, such as plover birds and bluestreak wrasse, so that participants would be unlikely to know the correct answers and would have to guess. In reality, half the animals exhibited the target behavior and half did not. Overall, participants estimated that cooperative behaviors are more common than competitive ones, and this asymmetry was larger for intraspecific behaviors (where the average difference was 10 percent) than for the interspecific behaviors (where the average difference was only 2 percent).

To measure participants' understanding of geologic time, we adapted a task from Lee et al. (2011). Participants were presented with eighteen historic or geologic events and were asked to estimate how much time had passed since the event occurred. They registered their estimate by selecting one of ten time periods, beginning with "between 100 and 1000 years ago" and ending with "between 100,000,000,000 and 1,000,000,000,000 years ago." The events included the time since Rome was founded, the time since the extinction of dinosaurs, the time since the Earth was formed, and the time since the Milky Way galaxy was formed. Consistent with the findings of Lee et al. (2011), participants systematically overestimated how much time had passed for events occurring less than 10,000 years ago and systematically underestimated how much time had passed for events occurring more than 10,000 years ago.

In sum, participants underestimated the prevalence of intraspecific variation, the prevalence of competition relative to cooperation (especially within a species), and the duration of geologic events. Still, participants varied in their accuracy on each task, and we ran a regression analysis to determine whether understanding each target concept relates to understanding evolution. We regressed scores on our measure of evolution understanding against scores on the intraspecific variation task (the proportion of traits judged potentially variable), scores on the intraspecific competition task (the proportion of behaviors accurately attributed), and scores on the geologic time task (the proportion of events accurately time-stamped). We used a stepwise regression, in which the predictor variables are entered into the regression model by the amount of variance they explain. The first predictor entered was intraspecific competition, which explained 15 percent of the variance in evolution understanding. The second was intraspecific variation, which explained an additional 4 percent. And the third was geologic time, which explained an additional 2 percent. All predictors were significant.

These results confirm the general finding that theory development in the history of science often parallels conceptual development in the individual. Just as Darwin's discovery of natural selection appears to have been based on the conceptual foundations of intraspecific variation, intraspecific competition, and geologic time, students' understanding of natural selection is based on the same foundations. That said, the relative contributions of these foundations were strikingly different. Appreciating within-species competition explained nearly four times as much variance as appreciating within-species variation and nearly eight times as much as appreciating geologic time. Thus, the focus on variation in the philosophy of science does not align with the psychology of evolution understanding. Recognizing that conspecifics compete for resources appears to be more critical to learning about evolution than recognizing that conspecifics vary in their traits.

From an empirical point of view, it's debatable whether organisms are truly more competitive than cooperative—that is, whether nature is better characterized as a "peaceable kingdom" or as "red in tooth and claw" (see De Waal, 2006). Regardless, the latter appears to foster a more accurate, population-based view of evolution. Indeed, what predicted participants' understanding of evolution was not their recognition of competition in general but their recognition of competition within a species. The better participants appreciated that members of the same species compete for resources, the better they understood the logic of natural selection and its consequences for phenomena as diverse as speciation and extinction.

Implications for understanding and improving scientific knowledge

Scientists' pathways to scientific truths are often quite different than students' pathways to the same truths. For instance, in trying to understand where species came from and why they are adapted to their environment, early biologists struggled with (a) the need to account for these phenomena within a naturalistic framework; (b) the need to account for a wide diversity of relevant data, from fossils to analogous traits to homologous traits; and (c) the deep-seated assumption that species are homogenous "types" rather than continuums of variation. Biology students, on the other hand, struggle with (a) finding value in a naturalistic explanation for phenomena they can already explain by divine creation, (b) interpreting ambiguous or misleading information about evolution conveyed through public discourse and public representations, and (c) conceiving of species as more competitive than cooperative—that is, recognizing that conspecifics compete for food, shelter, and mates. Differences like these are common in other domains of knowledge as well (noted earlier), and they likely have implications for the acquisition and representation of scientific concepts.

One implication is that different forms of cultural input can catalyze the same theory change—from an intuitive theory to a scientific theory—but the output of that process might not be the same, even if the starting point is the same. Humans typically converge on the same intuitive theories, despite living in different environments or in different time periods (Eckstein and Kozhevnikov, 1997; McCloskey, 1983; Vosniadou, 1994b; Wiser and Carey, 1983). Our innate ideas about objects, agents, and organisms furnish us with shared expectations about how those entities will behave, and those expectations are further refined through shared experiences (Carey, 2009; Spelke, 2000). For instance, early-emerging expectations about contact causality and free fall lay the groundwork for an intuitive theory of motion that varies little from one country to the next, whether it be China, Mexico, Israel, Turkey, Ukraine, or the Philippines (Shtulman, 2017).

Science can reshape and restructure our theories to a point where they are no longer intuitive, but it's an open question whether the now-counterintuitive theories are equally counterintuitive for those who discovered them as for those who learned them secondhand. The steps involved in deriving a scientific theory, via data and inference, may be critical to integrating that theory with the expectations and experiences that predated it. On the other hand, deriving a scientific theory could lead one to quarantine the theory, viewing it as relevant to controlled, lab-based observations but irrelevant to observations from everyday life. Scientific innovation is cultural innovation writ large, and there is much we still do not understand about how knowledge obtained through culture is combined with knowledge obtained through experience.

From a practical point of view, comparing scientists' and students' understanding of the same ideas can lead to more effective science education. One of the hallmarks of intuitive theories is their resistance to counterevidence and counterinstruction. Intuitive theories of evolution, for instance, have been documented in individuals of all levels of education, including college biology majors (Nehm and Reilly, 2007), medical-school students (Brumby, 1984), preservice biology teachers (Deniz, Donelly, and Yilmaz, 2008), and even graduate students in the biological sciences (Gregory and Ellis, 2009). Understanding evolution does not increase linearly with exposure to evolutionary ideas; a whole semester of college-level biology typically has no impact on a student's ability to grasp the logic of natural selection (Shtulman and Calabi, 2013).

One reason that instruction may fail to facilitate conceptual change is that it targets the wrong preconceptions. Instruction that follows the sequence of findings that led to the discovery of a scientific idea may miss the mark for nonscientists, who hold a different set of assumptions and struggle with a different set of concerns. Curricula informed by the history of science have proven effective in some domains (see, e.g., Wandersee, 1986), but they may not be effective in all domains or for all students. Additional research comparing the conceptual ecologies of scientists and students is needed to determine whether students should be led to scientific truths along the same path they were discovered or along different paths.

Conclusion

History repeats itself. Students of science today face many of the same difficulties in understanding scientific ideas as the scientists who discovered those ideas. The first theory of a domain explicitly articulated by scientists often resembles the intuitive theories implicitly constructed by nonscientists. That said, there is more than one pathway from intuitive theories to scientific theories, and the pathways taken by scientists may differ systematically from those taken by students. Here, I have outlined three factors that lead scientists and students down different paths: scientists and students hold different explanatory goals; they know of different empirical phenomena; and their theorizing is constrained by different background assumptions. These factors may alter how scientific ideas are mentally represented, either in relation to the world or in relation to each other, and they point to the need for additional research comparing the concepts and theories of professional scientists to those of science students. Scientific knowledge is instantiated in many forms—papers, models, technologies, the records of early scientists, the minds of science students—and all forms can shed light on the structure and origin of such knowledge, particularly if analyzed together.

References

Au, T. K. F., Chan, C. K., Chan, T. K., Cheung, M. W., Ho, J. Y., and Ip, G. W. (2008). Folkbiology meets microbiology: A study of conceptual and behavioral change. *Cognitive Psychology*, 57: 1–19.
Barrett, H. C. and Broesch, J. (2012). Prepared social learning about dangerous animals in children. *Evolution and Human Behavior*, 33: 499–508.
Blacker, K. A. and LoBue, V. (2016). Behavioral avoidance of contagion in children. *Journal of Experimental Child Psychology*, 143: 162–70.

Blancke, S., De Smedt, J., De Cruz, H., Boudry, M., and Braeckman, J. (2012). The implications of the cognitive sciences for the relation between religion and science education: The case of evolutionary theory. *Science & Education*, 21: 1167–84.

Blown, E. J. and Bryce, T. G. K. (2013). Thought-experiments about gravity in the history of science and in research into children's thinking. *Science & Education*, 22: 419–81.

Bonawitz, E. B., van Schijndel, T. J., Friel, D., and Schulz, L. (2012). Children balance theories and evidence in exploration, explanation, and learning. *Cognitive Psychology*, 64: 215–34.

Bowler, P. J. (1992). *The Eclipse of Darwinism: Anti-Darwinian Evolution Theories in the Decades around 1900*. Baltimore: John Hopkins University Press.

Brumby, M. N. (1984). Misconceptions about the concept of natural selection by medical biology students. *Science Education*, 68: 493–503.

Carey, S. (2009). *The Origin of Concepts*. Oxford: Oxford University Press.

Carey, S. and Spelke, E. (1996). Science and core knowledge. *Philosophy of Science*, 63: 515–33.

Catley, K. M. and Novick, L. R. (2008). Seeing the wood for the trees: An analysis of evolutionary diagrams in biology textbooks. *BioScience*, 58: 976–87.

Chi, M. T. H. (2005). Commonsense conceptions of emergent processes: Why some misconceptions are robust. *The Journal of the Learning Sciences*, 14: 161–99.

Chi, M. T. H., Roscoe, R. D., Slotta, J. D., Roy, M., and Chase, C. C. (2012). Misconceived causal explanations for emergent processes. *Cognitive Science*, 36: 1–61.

Chinn, C. A. and Brewer, W. F. (1998). An empirical test of a taxonomy of responses to anomalous data in science. *Journal of Research in Science Teaching*, 35: 623–54.

De Waal, F. (2006). Morally evolved: Primate social instincts, human morality, and the rise and fall of "Veneer Theory". In S. Macedo and J. Ober (Eds.), *Primates and Philosophers: How Morality Evolved* (pp. 1–58). Princeton: Princeton University Press.

Deniz, H., Donelly, L. A., and Yilmaz, I. (2008). Exploring the factors related to acceptance of evolutionary theory among Turkish preservice biology teachers. *Journal of Research in Science Teaching*, 45: 420–43.

DiSessa, A. A. (1993). Toward an epistemology of physics. *Cognition and Instruction*, 10: 105–225.

DiSessa, A. A. (2008). A bird's-eye view of the "pieces" vs. "coherence" controversy (from the "pieces" side of the fence). In S. Vosniadou (Ed.), *International Handbook of Research on Conceptual Change* (pp. 35–60). New York: Routledge.

Eckstein, S. G. and Kozhevnikov, M. (1997). Parallelism in the development of children's ideas and the historical development of projectile motion theories. *International Journal of Science Education*, 19: 1057–73.

Erickson, G. L. (1979). Children's conceptions of heat and temperature. *Science Education*, 63: 221–30.

Evans, E. M. (2001). Cognitive and contextual factors in the emergence of diverse belief systems: Creation versus evolution. *Cognitive Psychology*, 42: 217–66.

Evans, E. M., Spiegel, A. N., Gram, W., Frazier, B. N., Tare, M., Thompson, S., and Diamond, J. (2010). A conceptual guide to natural history museum visitors' understanding of evolution. *Journal of Research in Science Teaching*, 47: 326–53.

Fox, R. (1971). *The Caloric Theory of Gases: From Lavoisier to Regnault*. Oxford: Clarendon.

Gelman, S. A. (2003). *The Essential Child*. Oxford: Oxford University Press.

Gelman, S. A. and Legare, C. H. (2011). Concepts and folk theories. *Annual Review of Anthropology*, 40: 379–98.

Gopnik, A. and Wellman, H. M. (2012). Reconstructing constructivism: Causal models, Bayesian learning mechanisms, and the theory theory. *Psychological Bulletin*, 138: 1085–108.

Gould, S. J. (1992). *Ever Since Darwin: Reflections in Natural History*. New York: Norton.

Gould, S. J. (1996). *Full House: The Spread of Excellence from Plato to Darwin*. New York: Three Rivers Press.

Gregg, V. R., Winer, G. A., Cottrell, J. E., Hedman, K. E., and Fournier, J. S. (2001). The persistence of a misconception about vision after educational interventions. *Psychonomic Bulletin & Review*, 8: 622–26.

Gregory, T. R. and Ellis, C. A. J. (2009). Conceptions of evolution among science graduate students. *Bioscience*, 59: 792–99.

Gruber, H. E. (1981). *Darwin on Man: A Psychological Study of Scientific Creativity*. Chicago: University of Chicago Press.

Harlow, D. B., Swanson, L. H., Nylund-Gibson, K., and Truxler, A. (2011). Using latent class analysis to analyze children's responses to the question, "What is a day?" *Science Education*, 95: 477–96.

Heddy, B. C. and Nadelson, L. S. (2012). A global perspective of the variables associated with acceptance of evolution. *Evolution: Education and Outreach*, 5: 412–18.

Holton, G. J. (1988). *Thematic Origins of Scientific Thought: Kepler to Einstein*. Boston: Harvard University Press.

Hull, D. (1965). The effect of essentialism on taxonomy: 2000 years of stasis. *British Journal for the Philosophy of Science*, 15–16: 314–26.

Inagaki, K. and Hatano, G. (2004). Vitalistic causality in young children's naïve biology. *Trends in Cognitive Sciences*, 8: 356–62.

Johnson, S. C. and Solomon, G. E. (1997). Why dogs have puppies and cats have kittens: The role of birth in young children's understanding of biological origins. *Child Development*, 68: 404–19.

Kaiser, M. K., Proffitt, D. R., and McCloskey, M. (1985). The development of beliefs about falling objects. *Perception & Psychophvsics*, 38: 533–39.

Kalish, C. W. (1996). Preschoolers' understanding of germs as invisible mechanisms. *Cognitive Development*, 11: 83–106.

Kim, E. and Pak, S. J. (2002). Students do not overcome conceptual difficulties after solving 1000 traditional problems. *American Journal of Physics*, 70: 759–65.

Koslowski, B., Marasia, J., Chelenza, M., and Dublin, R. (2008). Information becomes evidence when an explanation can incorporate it into a causal framework. *Cognitive Development*, 23: 472–87.

Kuhn, T. S. (1962). *The Structure of Scientific Revolutions*. Chicago: University of Chicago Press.

Kuhn, D. (1989). Children and adults as intuitive scientists. *Psychological Review*, 96: 674–89.

Kuhn, D. and Pease, M. (2008). What needs to develop in the development of inquiry skills? *Cognition and Instruction*, 26: 512–59.

Lack, D. (1947/1983). *Darwin's Finches*. Cambridge: Cambridge University Press.

Laudan, L., Donovan, A., Laudan, R., Barker, P., Brown, H., Leplin, J., Thagard, P., and Wykstra, S. (1986). Scientific change: Philosophical models and historical research. *Synthese*, 69: 141–223.

Lederberg, J. (2000). Infectious history. *Science*, 288: 287–93.

Lee, H. S., Liu, O. L., Price, C. A., and Kendall, A. L. (2011). College students' temporal-magnitude recognition ability associated with durations of scientific changes. *Journal of Research in Science Teaching*, 48: 317–35.

Legare, C. H. and Shtulman, A. (2018). Explanatory pluralism across cultures and development. In J. Proust and M. Fortier (Eds.), *Interdisciplinary Approaches to Metacognitive Diversity* (pp. 415–32). Oxford: Oxford University Press.

Libarkin, J. C. and Anderson, S. W. (2005). Assessment of learning in entry-level geoscience courses: Results from the Geoscience Concept Inventory. *Journal of Geoscience Education*, 53: 394–401.

Libarkin, J. C., Anderson, S. W., Dahl, J., Beilfuss, M., Boone, W., and Kurdziel, J. P. (2005). Qualitative analysis of college students' ideas about the earth: Interviews and open-ended questionnaires. *Journal of Geoscience Education*, 53: 17–26.

Lorch, R. F., Lorch, E. P., Calderhead, W. J., Dunlap, E. E., Hodell, E. C., and Freer, B. D. (2010). Learning the control of variables strategy in higher and lower achieving classrooms: Contributions of explicit instruction and experimentation. *Journal of Educational Psychology*, 102: 90–101.

MacDonald, T. and Wiley, E. O. (2012). Communicating phylogeny: Evolutionary tree diagrams in museums. *Evolution: Education and Outreach*, 5: 14–28.

Marques, L. and Thompson, D. (1997). Misconceptions and conceptual changes concerning continental drift and plate tectonics among Portuguese students aged 16–17. *Research in Science & Technological Education*, 15: 195–222.

Mayr, E. (1982). *The Growth of Biological thought: Diversity, Evolution, and Inheritance*. Cambridge: Harvard University Press.

McCloskey, M. (1983). Naïve theories of motion. In D. Gentner and A. Stevens (Eds.), *Mental Models* (pp. 299–324). Hillsdale: Erlbaum.

Millman, A. B. and Smith, C. L. (1997). Darwin's use of analogical reasoning in theory construction. *Metaphor and Symbol*, 12: 159–87.

Morris, B. J. and Masnick, A. M. (2014). Comparing data sets: Implicit summaries of the statistical properties of number sets. *Cognitive Science*, 39: 156–70.

Morris, S. C., Taplin, J. E., and Gelman, S. A. (2000). Vitalism in naïve biological thinking. *Developmental Psychology*, 36: 582–95.

National Science Board (2014). *Science and Engineering Indicators*. Arlington: National Science Foundation.

Nehm, R. H. and Reilly, L. (2007). Biology majors' knowledge and misconceptions of natural selection. *BioScience*, 57: 263–72.

Nersessian, N. J. (1989). Conceptual change in science and in science education. *Synthese*, 80: 163–83.

Nettle, D. (2010). Understanding of evolution may be improved by thinking about people. *Evolutionary Psychology*, 8: 205–28.

Newport, F. (2010). Four in 10 Americans believe in strict creationism. Washington: Gallup Organization.

Oreskes, N. (1999). *The Rejection of Continental Drift: Theory and Method in American Earth Science*. Oxford: Oxford University Press.

Özkan, Ö., Tekkaya, C., and Geban, Ö. (2004). Facilitating conceptual change in students' understanding of ecological concepts. *Journal of Science Education and Technology*, 13: 95–105.

Pew Research Center (2015). *Public and Scientists' Views on Science and Society*. Washington: Author.

Reiner, M., Slotta, J. D., Chi, M. T. H., and Resnick, L. B. (2000). Naïve physics reasoning: A commitment to substance-based conceptions. *Cognition and Instruction*, 18: 1–34.

Samarapungavan, A. (1992). Children's judgments in theory choice tasks: Scientific rationality in childhood. *Cognition*, 45: 1–32.

Schauble, L. (1996). The development of scientific reasoning in knowledge-rich contexts. *Developmental Psychology*, 32: 102–19.

Shtulman, A. (2006). Qualitative differences between naïve and scientific theories of evolution. *Cognitive Psychology*, 52: 170–94.

Shtulman, A. (2014). *Using the History of Science to Identify Conceptual Prerequisites to Understanding Evolution*. Poster presented at the 40th meeting of the Society for Philosophy and Psychology, Vancouver, Canada.

Shtulman, A. (2017). *Scienceblind: Why Our Intuitive Theories about the World Are So Often Wrong*. New York: Basic Books.

Shtulman, A. and Calabi, P. (2013). Tuition vs. intuition: Effects of instruction on naïve theories of evolution. *Merrill-Palmer Quarterly*, 59: 141–67.

Shtulman, A. and Checa, I. (2012). Parent–child conversations about evolution in the context of an interactive museum display. *International Electronic Journal of Elementary Education*, 5: 27–46.

Shtulman, A., Neal, C., and Lindquist, G. (2016). Children's ability to learn evolutionary explanations for biological adaptation. *Early Education and Development*, 27: 1222–36.

Shtulman, A. and Schulz, L. (2008). The relationship between essentialist beliefs and evolutionary reasoning. *Cognitive Science*, 32: 1049–62.

Smith, C. L. (2007). Bootstrapping processes in the development of students' commonsense matter theories: Using analogical mappings, thought experiments, and learning to measure to promote conceptual restructuring. *Cognition and Instruction*, 25: 337–98.

Smith, C., Carey, S., and Wiser, M. (1985). On differentiation: A case study of the development of the concepts of size, weight, and density. *Cognition*, 21: 177–237.

Sober, E. (Ed.) (1994). *Conceptual Issues in Evolutionary Biology*. Boston: MIT Press.

Songer, C. J. and Mintzes, J. J. (1994). Understanding cellular respiration: An analysis of conceptual change in college biology. *Journal of Research in Science Teaching*, 31: 621–37.

Spelke, E. S. (2000). Core knowledge. *American Psychologist*, 55: 1233–43.

Springer, K. and Keil, F. C. (1989). On the development of biologically specific beliefs: The case of inheritance. *Child Development*, 60: 637–48.

Thagard, P. R. (1978). The best explanation: Criteria for theory choice. *The Journal of Philosophy*, 75: 76–92.

Thagard, P. (1999). *How Scientists Explain Disease*. Princeton: Princeton University Press.

Toulmin, S. and Goodfield, J. (1961). *The Fabric of the Heavens*. Chicago: University of Chicago Press.

Toulmin, S. and Goodfield, J. (1962). *The Architecture of Matter*. Chicago: University of Chicago Press.

Trend, R. (2001). Deep time framework: A preliminary study of U.K. primary teachers' conception of geological time and perceptions of geoscience. *Journal of Research in Science Teaching*, 38: 191–221.

Vosniadou, S. (1994a). Capturing and modeling the process of conceptual change. *Learning and Instruction*, 4: 45–69.

Vosniadou, S. (1994b). Universal and culture-specific properties of children's mental models of the earth. In L. A. Hirschfeld and S. A. Gelman (Eds.), *Mapping the Mind: Domain Specificity in Cognition and Culture* (pp. 412–30). Cambridge: Cambridge University Press.

Vosniadou, S. and Brewer, W. F. (1994). Mental models of the day/night cycle. *Cognitive Science*, 18: 123–83.

Wandersee, J. H. (1986). Can the history of science help science educators anticipate students' misconceptions? *Journal of Research in Science Teaching*, 23: 581–97.

Wertz, A. E. and Wynn, K. (2014). Selective social learning of plant edibility in 6- and 18-month-old infants. *Psychological Science*, 25: 874–82.

Wiser, M. and Amin, T. (2001). "Is heat hot?" Inducing conceptual change by integrating everyday and scientific perspectives on thermal phenomena. *Learning and Instruction*, 11: 331–55.

Wiser, M. and Carey, S. (1983). When heat and temperature were one. In D. Gentner and A. L. Stevens (Eds.), *Mental Models* (pp. 267–97). Hillsdale: Erlbaum.

Zimmerman, C. and Cuddington, K. (2007). Ambiguous, circular and polysemous: Students' definitions of the "balance of nature" metaphor. *Public Understanding of Science*, 16: 393–406.

6

Intuitive Epistemology: Children's Theory of Evidence

Mark Fedyk, Tamar Kushnir, and Fei Xu

Introduction

This chapter is premised upon a simple but potentially powerful assumption: it is not possible for the mind to acquire much of its knowledge without it also possessing an intuitive understanding of a set of epistemological concepts—namely, whatever concepts are just those which can be used by the mind to identify, in the stream of information that comes from the world, properties relevant to the formation of accurate belief. It is likely that extremely simple concepts of probability, causation, and testimony, along with the concepts needed to form very basic epistemic, statistical, and logical generalizations, are among the first members of the relevant set. Perhaps an elementary concept of knowledge is necessary too. But whatever the exact or initial membership of this set is, we believe that positing such a set is necessary to explain the fact that, even in young children, making inferences about the sorts of things which are relevant to accurate beliefs leads, often enough, to learning. Accordingly, we shall call the conceptual resources contained in this set, and which facilitate the formation of accurate belief, a learner's *intuitive epistemology*.

Our aim in this chapter is to develop support for a thesis that is a corollary of the view that the mind has an intuitive epistemology. We will argue that a learner's intuitive epistemology includes a *theory of evidence* by approximately the age of four. What this means is that the child has a grasp of enough epistemic concepts and interlinking principles to make, frequently enough, accurate judgments about what sorts of events, effects, and occurrences do and do not count as evidence. Additionally, we believe that the relevant concepts for evidence are not encoded or represented in the mind in an unstructured fashion—the concepts for evidence are not more or less randomly distributed across (or among) the child's other concepts, prototheories, theories, and various other forms of mental representation. Our view, instead, is that the child's concepts of evidence are encoded as a *theory*, which means that the concepts are embedded within a web of inferential connections that hold, primarily, among the evidential concepts themselves, but also, secondarily, between the evidential concepts and a set of additional nonevidential concepts. The existence of the inferential structure linking the child's evidential concepts with one another is why it is appropriate to speak of a child possessing a *theory* of evidence.

A theory of mind defines a specific domain of knowledge: the conceptual content of a theory of mind demarcates what any individual can and cannot treat as a mental state. But a theory of evidence is not a domain-specific theory in that sense. Why? The epistemic concepts a theory of evidence contains can be employed when constructing various other domains of knowledge; for example, an unreliable informant can potentially mess up a child's intuitive physics or theory of mind or intuitive biology. The conceptual content of a theory of evidence therefore intersects with most other (perhaps *all* other) domains that are individuated by the nature of their conceptual content. So, since a theory of evidence cannot define a *sui generis* domain of knowledge, it is best thought of as a domain-general theory.

It is also likely that a theory of evidence, like domain-specific theories such as a child's theory of mind or of number, can be refined as a byproduct of the child's more basic capacity to make general, abstract, and causal inferences about the structure of her world (Gopnik and Meltzoff, 1997). At the same time, it is improbable that the conceptual content of a theory of evidence is constructed entirely as a byproduct of learning. Children are probably endowed with a handful of rudimentary concepts for evidence from the day that they are born. Here, the argument is familiar: it is hard to explain how a child could begin to construct knowledge of various domains without a grasp of at least some very simple epistemic concepts. These concepts, however, may be refined—or even completely reconstructed—as learning and development subsequently occur.

So, our conjecture is that young children may begin learning equipped with a small number of elementary epistemic concepts. Then, by constructing and evaluating beliefs about the world at different levels of generalizations and abstraction and in relation to different kinds and forms of evidence, children come to form both more refined concepts of evidence and also acquire specific principles which link their concepts of evidence with other concepts and principles. In so doing, they begin to construct the earliest forms of their theory of evidence by constructing principles that link together their growing stock of concepts for evidence. Finally, after enough time, children seem to be capable of sophisticated reasoning about evidence in support of their learning—an observation which raises the interesting possibility that refining (and refining earlier refinements of) a theory of evidence is one of the ways a child learns to learn.

Here is how this chapter is organized. We will use the section titled "A theory of evidence: Theories support abstraction" to further clarify what it means to impute a theory of evidence to a young child. The section titled "Evidence of children's theory of evidence" then surveys some of the experimental evidence that we believe supports the existence of a psychological theory of evidence. Recent findings suggest that even very young children are surprisingly sensitive to different sources and types of evidence, and that this sensitivity informs their judgments in a way that is, frequently enough, conducive to the acquisition of knowledge.

Our focus changes in the section titled "Epistemologized psychology: Cognitive psychology as epistemology," where we pivot from a discussion of the empirical work which supports our contention that children have a theory of evidence to an examination of a new idea about how to implement the long-standing goal of naturalizing epistemology. It is surprising that, in light of the nearly overwhelming amount of philosophical scholarship on the question of how epistemology can be naturalized [cf. (Quine, 1969; Johnsen, 2005; Feldman, 2011)], one idea seems to be totally absent in the literature: that epistemology can be naturalized by *epistemologizing* cognitive psychology— where this means adopting as a working methodological idea the principle that normative concepts should be held to the same standard as nonnormative concepts throughout the formulation, testing, and acceptance or rejection of psychological theories. If a concept earns its place in a theory in the cognitive sciences by virtue of its ability to contribute to deep and oftentimes novel explanations, it should not matter whether that concept is (even a very thick) normative concept or not. Our view, then, is that such an *epistemologizing* of psychological methods will be among the effects of pursuing further research about either children's theory of evidence or, more generally, the mind's intuitive epistemology. It will be very hard to study how people learn about and use evidence without relying on deep commitments about what should and should not count as evidence—or, to put the same point another way, it will be very hard to study how people learn about and use evidence without using many of the epistemological concepts we mentioned above in the first paragraph. We further explore some of the philosophical implications of the idea of an epistemologized developmental psychology in the section "Epistemologized psychology: Cognitive psychology as epistemology," before turning to concluding remarks in the final section "Conclusion."

A theory of evidence: Theories support abstraction

The word *theory* can be used as a technical term in psychology: theories are imputed to a person as constituents of their individual psychology to explain both the person's thinking and the impact of thought on their judgments and behavior. To return to an example from above, a theory of mind is meant to explain mindreading, which is the capacity in a person to predict and understand the mental states of others. What it means to impute a theory of mind, specifically, is to say that a person has conceptual representations of several abstract concepts and principles, and that this person can apply these representations to another's cognition in order to understand, predict, and otherwise interact intentionally with this other person.

It will be helpful to unpack this example a bit more. A theory of mind that is useful for making predictions about people's future behaviors and for coming up with explanations of their past behaviors must include, at the very least, concepts for desires, intentions, preferences, psychological causes, and beliefs. Furthermore, among the principles likely included in a minimally useful theory of mind are principles such as <if person P desires outcome X, P believes that doing Y will cause X, and P is in a position to do Y, then P will do Y>; <If P was observed doing Y, then P desired either Y or something else, Z, that was a direct causal effect of doing Y>; and <If person P is reaching her hand toward object O, it is because P wants O>. Crucially, these principles show how theories can create inferential connections between concepts that refer to observable things (an open hand) and concepts that refer to unobservable things (psychological desire)—so, these principles also provide a straightforward illustration of how one of the cognitive functions of theories is to confer the capacity to reason about things that are, either literally or metaphorically, beyond the limits of direct sensory detection. The usefulness of a theory can, therefore, be a function of the richness of the abstract concepts found within a theory and the capacity of the theory to encode a meaningful number of inferential connections between those abstract concepts and concepts that refer to more concrete (observable, or easily perceptible) objects, kinds, and causes. Put more simply: theories are one way that we can make inferences about things, kinds, objects, processes, and properties that cannot be directly perceived.

We believe that the available experimental evidence suggests that it may be worthwhile exploring the hypothesis that children have a theory of evidence that is, in several important ways, analogous to their theory of mind. From birth, if not earlier, all children are the recipients of a constant stream of information from the world, their bodies, and the people around them. Only a small portion of this information is relevant to learning; much of this river of data can be discarded or ignored without any impairment to a child's subsequent learning. However, as we just noted, it is rarely if ever perceptually obvious which items of information are relevant to, or useful for, the formation of accurate belief; the world does not automatically place a label "this is pertinent to learning; treat this as evidence" on only the items of information that can facilitate the formation of accurate beliefs about the world. Classifying information as evidence requires abstraction, and abstraction can be facilitated by a mental theory. Accordingly, we suggest that one of the tools that learners use to selectively classify information that they are receiving from the world as evidence is a theory of evidence, because just such a theory can encode the principles which drive the ability to make inferential connections between perceptually salient effects, events, objects, and properties[1] and the members of a set of (abstract) concepts that each refer to different types, kinds, instances, and forms of evidence.

A theory of evidence: Theories link judgments with contexts and goals

But that is not all that mental theories do. A further reason why it is necessary to posit a *theory* of evidence, and not merely knowledge of an unstructured set of abstract concepts for evidence, is that what information counts as evidence for a learning depends upon at least two additional factors: the contexts in which different kinds of learning both are and are not possible, and the learning goals or outcomes that the child can possibly pursue.

The reason that a child must be able to reason about how context and goals interact with the judgments by which she classifies information as evidence is that no information counts as evidence in a categorical or absolute sense. For example, and to foreshadow our discussion of the scientific research which supports our conjecture, the set of the kinds of evidence that can be used to learn the meaning of words has very little overlap with the set of the kinds evidence that can be used to learn the function of artifacts like hammers, balloons, or wheels—so, the learning goal (learning about words vs. learning about causal functions) places a constraint on the kinds of evidence that someone should be on the lookout for. Likewise, whether one and the same bit of information should be treated as evidence can vary from context to context. If a parent asserts that "that is a blicket" while idly daydreaming in the presence of a child who is playing in the same room but nowhere near the parent, the parent provides the child with no evidence whatsoever. But if, while at a science museum, a parent points at a brightly colored box that is adorned with flashing lights and, while making direct eye contact with a child, asserts "that is a blicket," then the child receives information that is appropriately classified as evidence. So, a child learning about blickets, for example, must understand that, in some contexts but not others, assertions about blickets can, and should, be treated as evidence.[2]

Importantly, the relationship between contexts, goals, and types of evidence is not fixed and unmalleable. As learning imbues the mind with increasingly complex knowledge structures, the relationship between evidence, goals, and context can change dramatically. A person who learns, for instance, some of the rules of deductive logic thereby learns principles that can, inter alia, be used to classify mutually exclusive events as a type of evidence, a development which dramatically increases the number of contexts in which causal learning can occur. Likewise, someone who learns about the character trait of honesty learns about a kind of evidence that also expands the number of learning goals a learner can work toward achieving—as testimony is one of the most powerful drivers of learning (Tomasello, 2014; Lackey, 2008; Stephens and Koenig, 2015; Koenig and Harris, 2007; Koenig, Clément, and Harris, 2004). What's more, learning about honesty also expands the number of contexts in which interpersonal learning both can and cannot occur.

Thus, a person's theory of evidence can be thought of as the conceptual tools they use to make judgments both about *what* information counts as evidence (this requires abstraction) as well as *where* and *when* a learner should and should not be looking out for the kinds of evidence that she knows about (this requires knowledge of learning goals and contexts). Or, to put these ideas even more explicitly, our hypothesis is that a theory of evidence includes

(a) A set of abstract concepts, which probably includes simple concepts of probability, causation, and testimony, along with the concepts needed

to form very basic epistemic, statistical, and logical principles and generalizations;
(b) Inferential principles connecting the concepts of evidence with other nonabstract concepts;
(c) Inferential principles connecting concepts of evidence with concepts referring to different types of learning contexts and different types of learning goals—for example, learning from adults versus learning from one's peers, learning about the meanings of words versus learning about the rules of the game, and so on.

It is our view that very simple concepts of evidence, learning goals, and contexts populate the mind's earliest instantiations of a theory of evidence. But because a theory of evidence is itself a byproduct of learning, it is entirely possible—and, as the examples we used above suggest, we believe quite likely—that a person's theory of evidence undergoes substantial increases in its richness and complexity over the course of their own cognitive development.

Evidence of children's theory of evidence

That said, our focus in this chapter is only the theory of evidence as it likely exists in the minds of children at about the age of four. And, the scientific argument that children rely upon a theory of evidence in order to acquire some of their knowledge is straightforward. If children have a theory of evidence roughly as we have defined it above, then, on abductive grounds, we should observe children

1. Making judgments in which they treat different kinds of information *as if* it is evidence; and where
2. These judgments are usually *context* and *goal* appropriate; and where, partially because of this,
3. The information that is treated *as if* it is evidence is information that *should be* treated as evidence; and where, because of (1), (2), and (3),
4. The judgments are usually supportive of learning.

We will now describe five cognitive abilities that children use to facilitate their learning, each of which looks like it satisfies our quadripartite prediction. There are more examples in the literature which also fit our prediction—however, the following are the clearest examples that we know of, and by limiting ourselves to a discussion of just the following five capacities, we can keep this chapter reasonably focused.

Accuracy monitoring

As a rule of thumb, any indication that some information is accurate is indication that the information is *potentially* evidence. Consistent with this idea, recent work using studies of how children learn from the people they are interacting with has shown that children rely on a number of proxies for accuracy of information. Perhaps the simplest example of this occurs when a prior history of accuracy in labeling objects that the child is familiar with is taken by the child to mean that the person doing the labeling is a reliable speaker. Children then project this estimation of reliability by trusting new labels introduced by the same person for both novel words and novel object functions (Birch, Vauthier, and Bloom, 2008; Koenig and Harris, 2005). Children are vigilant monitors of the content of people's speech to them, checking what they are hearing for consistency and conflict, but they also monitor and track variations in speakers' moral behavior, mutual consensus, and group membership (Hetherington, Hendrickson, and Koenig, 2014; Mascaro and Sperber, 2009; Corriveau, Fusaro, and Harris, 2009)—these are all properties that can be reliably interpreted by learners as "proxies" for accuracy, even though each differs in how complex the association between accuracy and the proxy is likely to most frequently be. Indeed, by age four, children can monitor the probability of accuracy; they are able to decide how likely someone is being reliable, as opposed to making simple deterministic "yes/no" judgments of a person's reliability (Pasquini et al., 2007). Children are also more likely to trust speakers who are members of their linguistic community—as evidenced by, for instance, speaking with the same or a familiar accent (Kinzler, Corriveau, and Harris, 2011). Since children must learn many of the finer details of the local social worlds that they inhabit, and since these social worlds are constructed, in part, by the linguistic practices of their inhabitants, it is rational to accord speakers who are members of the same linguistic community a higher degree of trust. Finally, we can see how some of the complexity of children's theory of evidence by looking at studies which require children to compare two different potentially accurate sources of evidence. Children frequently use age (adults vs. other children) as a proxy for accuracy, but this can be overridden by independent observations of specific instances of accuracy or reliability, such as when a young child knows more than an adult about, for instance, a character in a story or a movie (Jaswal and Neely, 2006).

Recognizing knowledge and distinguishing between knowledge and ignorance

But children do not only rely upon a variety of proxies for accuracy—they are also able to reason, much more directly, about knowledge and ignorance. Of course, one person's knowledge should be another person's evidence. In line with this principle, and even from very early in development, children are able to distinguish between people, including themselves, according to knowledgeability. At the age of twelve months, infants have been shown to point more to the location of an object when they see an adult who is ignorant of the object's location looking for the object, compared with an adult who has knowledge of the object's precise location (Behne et al., 2012; Liszkowski, Carpenter, and Tomasello, 2008; O'Neill, 1996). By the age of sixteen months, infants use pointing gestures as interrogative demands—in order to elicit information—but only from people who are knowledgeable of the relevant information (Begus and Southgate, 2012; Southgate, van Maanen, and Csibra, 2007).

It is therefore hardly a surprise that, by the age of two, children can offer verbal reports of their own knowledge and ignorance, calibrate these reports in degrees of certainty, and modulate or refine these reports in light of self-observation (Shatz, Wellman, and Silber, 1983; Furrow et al., 1992); or that, by the age of three, children are able to accept or reject claims that are, by an independent baseline, highly reliable. For example, children may choose to reject claims made when a speaker asserts their own ignorance or uncertainty about a specific claim that they have made; for example, "Hmm, I don't really know what this is but I think it is a blicket" (Sabbagh and Baldwin, 2001; Sabbagh and Shafman, 2009; Henderson and Sabbagh, 2010). And finally, in studies that ask children to interact with two informants who differ consistently in the knowledge that they profess to have, children display, systematically, a preference for agreeing with the informants who seem to have more knowledge over informants who profess their own ignorance, and over informants who make incorrect guesses but make their uncertainty clear, for example, "Hmm, I'm not sure. I'll guess it's read" (Mills et al., 2011). Children also make proactive choices based on their estimates of knowledgeability; by age three, children will direct more questions to knowledgeable people than people who are ignorant, and by age five, children will direct more questions to a knowledgeable person than a person who makes plausible but inaccurate guesses (Mills et al., 2011).

Causal learning is also influenced by children's ability to conceptualize their world as containing people who have varying degrees of knowledge. In a recent study, children were presented with one of two variables: whether a potential informant was knowledgeable or ignorant about a novel toy, and also whether the informant was permitted to use that knowledge in performing an action. Then, children observed, in all of the conditions in this study, the two informants performing identical causal actions—and where, importantly, the actions themselves were unconstrained, equally intentional, and equally strongly associated with the effect that they produced. Children were more likely to attribute causal efficacy to the informants who were knowledgeable than to those who were not, suggesting that estimates of causal efficacy depend, even at a very young age, on accurate judgments about knowledge possession (Kushnir, Wellman, and Gelman, 2008). Furthermore, a similar study shows that an informant's statement about knowledge or ignorance about the causal properties of a toy influence whether preschool-aged children will imitate the informant's actions faithfully or not (Buchsbaum et al., 2011).

Assessing relative expertise

Children can reliably distinguish between knowledge and ignorance. But they can also make reasonably sophisticated estimates of relative expertise—such as when they compare the accuracy of the knowledge of two otherwise knowledgeable, or at least not obviously ignorant, informants.

The simplest example of this comes from the various studies showing that, by about the age of four, children know that different people know different things (Lutz and Keil, 2002; Danovitch and Keil, 2004). For example, mechanics are more likely to help with fixing bikes, whereas biologists know more about bird migration. But children are also able to make projective inferences about what (additional) knowledge a person is likely to possess on the basis of learning about some of the knowledge the person has. Children were introduced to one person who knew when objects would activate a special machine in a certain way and another person who knew when objects would activate the machine in another way (Sobel and Corriveau, 2010). Children were then shown objects that possess one of these different causal properties and were asked to endorse one of the confederates' novel labels for the objects.

In a similar study, preschoolers were introduced to two informants, one of whom (the "labeler") properly named two tools but failed at fixing two broken tools, and the other of whom (the "fixer") did not know the name for the tools but was able to fix the broken toys (Kushnir, Wellman, and Gelman, 2013). Both three- and four-year-olds selectively directed requests for new labels to the labeler and directed requests that a toy be fixed to the fixer. Then, in a second experiment in this study, four-year-olds also endorsed the fixer's causal explanations for the toy's mechanical failures, but not also the fixer's new names for objects. Together, these findings suggest that young children are able to represent both the scope and limit of other people's expertise.

Estimations of the relevancy of information to learning

Related to these judgments about the expertise of potential informants, children are also able to make sound inferences about whether or not novel information is relevant to the learning at hand. Evidence of this comes from studies of preschoolers that examine how pedagogical (or "ostensive") cues such as eye contact, child-directed speech, and generic language might provide a signal to children that information is being "taught" to them, and also for some epistemically meaningful purpose. For instance, children generalize further (Butler and Markman, 2012), imitate more faithfully (Southgate, van Maanen, and Csibra, 2009; Brugger et al., 2007), and restrict exploration (Bonawitz et al., 2011) as an effect of observing actions that are pedagogically demonstrated.

Relevancy is perhaps the simplest way of assessing whether novel information should be treated as evidence; these studies, therefore, suggest that children have a useful concept, albeit simple, for one type evidence by the age of four. But furthermore, one explanation for why pedagogical demonstrations have an impact on children's reasoning is that they invite children to make inferences about the social or cultural relevance of actions (Moll and Tomasello, 2007; Southgate, van Maanen, and Csibra, 2009). If this is right, then the estimates of relevancy produced by interactions between by a child's theory of evidence and external pedagogical cues are not only assessments of mechanical or causal relevance; they are also assessments by which a child can come to learn important facts about how her social worlds can facilitate learning. In light of this, it has been suggested that these interactions are a key driver of some of the impressive patterns of cultural learning that seem to be proprietary to our species (Csibra and Gergely, 2009, 2011).

Pedagogical cues have also been shown to support learning about both abstract categorization and object function. Several studies have demonstrated that children use pedagogical cues as an indication of which features of an object can be used to determine the object's correct categorization. The general method used throughout this work is one in which surface features of objects (such as color and shape) and nonobvious features (such as internal structure of causal powers that become apparent only after interactions with the object) are used to generate conflicting categorization judgments. When deciding between trusting categorizations based on surface features versus categorizations based on nonobvious features, preschoolers will, all things being equal, prefer the surface feature categorizations. However, after receiving particular pedagogical demonstrations of categorizations based upon nonobvious features, children are more likely to follow suit and use nonobvious features as the basis for their categorization judgments (Williamson, Jaswal, and Meltzoff, 2010; Butler and Markman, 2014; Yu and Kushnir, 2015). Of particular importance is the observation, in the last of these studies (Yu and Kushnir, 2015), that children show an equal interest in exploring the nonobvious features of objects whenever they are demonstrated—it is always fun to play with objects that make interesting sounds. Nevertheless, children in this study did distinguish between cases where nonobvious features are relevant to categorization and when they are not. Children's understanding of evidence for categorization, then, extends beyond their ability to distinguish between surface and nonobvious features.

In short, children are able to distinguish between when properties are relevant to learning and when they are not. Frequently, pedagogical cues assist them with this task; we are not claiming that a theory of evidence itself is sufficient to produce most forms of learning that are possible by the age of five. Rather, and again, the suggestion is that these cues interact with elements of a child's theory of evidence. The child's theory of evidence tells her that she is in a pedagogical context and that certain verbal cues are a source of evidence, while the cue itself provides the content of the evidence and may even, over time, lead to an enrichment of the theory of evidence itself so as to include concepts for the types of evidence that, earlier on, were the focus of pedagogical interactions.

Early attention to source of information

Finally, some indirect evidence of a theory of evidence comes from a cognitive ability that may precede the development of a psychological theory—the ability to simply scan and filter the near environment for information that can be treated as evidence. The existence of a capacity for what might be called *selective evidential filtering* is suggested by work showing that infants can ignore certain features of perceptual input and focus on the parts that are potentially relevant for making inferences based on probability (Denison and Xu, 2010, 2012; Xu and Denison, 2009).

To be clear, selective evidential filtering is not an example of an ability that seems well explained only by positing a theory of evidence. It is important to distinguish between making principled inferences about information that may or may not count as evidence, and, more simply, being able to attend to sources of information that could potentially be evidence. A theory of evidence is necessary to explain the former, while possession of a handful of mostly unconnected evidential concepts can account for the later. Nevertheless, evidence of selective evidential filtering is indirect evidence of a child's theory of evidence, simply because it would be very surprising if children made the leap from possessing no concepts of evidence whatsoever to possessing a network (i.e., a theory) of concepts of evidence and principles governing the use of those concepts in reasoning. What is more plausible is that there is an intermediary developmental stage, in which infants or very young children have and are able use concepts of evidence but do so only in ways that do not suggest that these concepts drive much in the way of deep inferences.

Still, it would be a mistake to think that all of children's ability to reason about evidence can be explained without postulating a *theory* of evidence. This is demonstrated by a recent study in which four-year-old children were given identical data that could help them learn words that functioned as simple labels for novel objects (Xu and Tenenbaum, 2007). Crucially, the data differed only in its source—whether it come from a knowledgeable teacher or the learner themselves. The children in this study learned the labels more accurately when this information was acquired from the teacher, seemingly indicating that, by about the age of four, children can integrate information about context with their judgments about evidence.

Epistemologized psychology: Cognitive psychology as epistemology

Stepping back now, what these studies show is that children can make rational inferences about evidence. Again, we suggest that a psychological theory of evidence is an attractive scientific explanation of children's ability to do this.

But the fact that a theory of evidence is itself a scientifically plausible explanation of some aspects of human cognition generates a further, and apparently novel, philosophical implication about what it means to naturalize epistemology. The conclusion we have just arrived at raises the intriguing possibility that experimental epistemology may be a subfield of cognitive psychology—that is, the subfield of cognitive psychology that employs intrinsically normative, intrinsically epistemological concepts and principles to study the epistemically relevant psychologically phenomena, such as learning, perception, memory, and testimony. Put another way, the fact that the hypothesis that children's rational learning may be facilitated by a theory of evidence is scientifically plausible provides us with further reason to think that it may be possible to naturalize epistemology by *epistemologizing* psychology.

Allow us to explain. First of all, it is important to stress that the idea of epistemologizing psychology is not the same idea as the frequently mooted Quinean dictum that epistemology can be naturalized by replacing it with a branch of psychology (Quine, 1969). As Quine's dictum is probably most frequently interpreted, it is taken to mean that an existing body of scientific research spanning psychophysics to cognitive psychology will be able to answer most of the traditional questions in epistemology—and, in so doing, this research will render the existing field of epistemology redundant. Yet, there is an ambiguity in the Quinean dictum that is easy to overlook if the dictum is understood at only this level of abstraction. The dictum, specifically, does not address the psychological/causal question of how much *rationality* is required to induce meaningful learning (Putnam, 1982, pp. 20–21). In the abstract, Quine's dictum is compatible with the proposition that the mind produces accurate beliefs by a series of entirely mechanical transformations performed on the information derived from the sensory transduction, with no normative or computational processing required at any point along the chain from sensation to belief. But an alternative view is that both normative and computational processing is necessary for the mind to construct a deep and rich network of mostly accurate beliefs on the basis of sensory experience—it is not possible to learn about the world without reasoning about it, and it is not possible to reason about the world without recruiting or implicating some normative concepts to the computational processes that constitute such reasoning. If the latter is the case, then it will not be possible for psychology to study how accurate beliefs are formed without embedding any number of intrinsically normative concepts in the formulation of (even just empirically adequate) psychological theories—because these concepts must be used in order for the relevant theories to be able to describe *how* it is that the mind does what, rationally speaking, it *should* do. And, since Quine published his famous article, one of the most significant historical lessons of almost four decades of research in cognitive science is that an *immense* amount of processing is required in order to turn sensory information into accurate belief (Johnson-Laird, 1988; Tenenbaum et al., 2011; Xu and Kushnir, 2013; Marr, 1982). As indicated, this fact means that scientific explanations of how reasoning facilitates the acquisition of accurate beliefs will depend upon the injection into psychological theories and methods of any number of (intrinsically normative) epistemic concepts—or, to put this conclusion another way, in order to discover, scientifically speaking, how accurate beliefs are formed, cognitive psychology must, to some important degree, be epistemologized.

There is a different path to the exactly the same conclusion. It is often asserted, though much less frequently argued, that the deepest methodological difference between philosophy and science is that philosophy is about either purely conceptual matters or purely normative matters, while science concerns itself almost exclusively with descriptive matters of empirical fact [cf. (Longino, 1996)]. Yet, if this pair of ideas were adopted as part of the methodological framework used by a cognitive scientist or a philosopher interested in studying the functional role that a theory of evidence plays in learning, the ideas would work together to block any scientific research. This is because one cannot formulate causal-explanatory hypotheses about the cognitive function of a psychological theory of evidence—or, more generally, an intuitive epistemology, or even knowledge of general statistical and logical principles—without making two kinds of commitments.

The first are commitments about what kinds of concepts the mind needs in order to be able learn by reasoning about such things as which should count as evidence. Since at least some, and probably most, of these concepts will be epistemological concepts and will thus be inherently normative concepts, there will be no way to formulate hypotheses meant to explain learning without making commitments that, at minimum, amount to the position that the minds of learners frequently use intrinsically normative concepts and that they, at least almost as frequently, use those normative concepts in the way that those concepts *should be* used.

The second commitment is more abstract. It is a commitment that comes in the form of methodological openness to the possibility that, as research into such topics as the mind's intuitive epistemology progresses, novel *sui generis* epistemological conclusions may emerge as byproducts of ordinary scientific inquiry [cf. (Xu, 2007, 2011; Fedyk and Xu, 2017)]. For example, in related work, we have argued that learners have *prima facie* right to the exercise of a complex ability that we call "cognitive agency" (Fedyk and Xu, 2017). But, obviously, to commit oneself to the proposition that science can make progress toward answering descriptive questions only is to deny this very possibility. So, the fact that a theory of evidence is a scientifically plausible hypothesis in developmental psychology shows, working backward through these inferences, that it would be a mistake to hold that there is methodological dissociation between philosophy and science such that philosophy is about conceptual or normative matters *only* and science about descriptive or empirical matters *only*. As research on children's theory of evidence demonstrates, progress in psychology science can sometimes depend essentially upon the normative content of its causal-explanatory theories.

These two lines of reasoning, thus, each lead us to exactly same conclusion—namely, that it can be inductively and explanatorily fruitful in cognitive psychology to use intrinsically normative-epistemological concepts, assumptions, hypotheses, and principles. By this, we do not just mean that sometimes it will be scientifically fruitful to use a particular normative concept in a (purely?) descriptive way—such as to simply characterize the content of someone's mental states. We mean, in addition to this, that, often enough, it may be scientifically fruitful to formulate psychological theories that are themselves normative; they describe some aspect of the cognitive system while also saying what cognitive systems should (or should not) do. And so, what the proposal that psychology be epistemologized amounts to is that we take up this latter reading of our conclusion and run with it as far as the science will allow. Pursuing an increase in the normative-epistemological content of psychology as far as the data will take us is what it means, methodologically speaking, to epistemologize psychology.

To return briefly to an idea broached a few paragraphs back, the coherence of epistemologized psychology shows that it is a mistake to hold the view that naturalizing epistemology must consist of a search for analytical concepts that can reduce epistemology to some preexisting research somewhere in the behavioral sciences—or even just searching the natural sciences for concepts and principles that seem like they can help resolve classical problems in epistemology, like the definition of knowledge (Kornblith, 2002). Instead, the most productive and integrative way to naturalize epistemology could be to work how to *elevate* research in psychology to the status of epistemology.

Furthermore, the coherence of epistemologized psychology also demonstrates that experimental epistemology may simply be contemporary cognitive psychology, albeit only after the latter is imbued with sufficient normative concepts and principles, and the latter also begins to operate from a set of methodological norms expanded to create the space to pursue scientific answers to questions like "How should I reason?" "When should I stop searching for new knowledge?" and "Which of my beliefs are most trustworthy?" Of course, there is no guarantee that the epistemological concepts which carry the most inductive or explanatory weight in even minimally epistemologized psychology will also be the highly refined, mostly technical concepts that are central to debates in contemporary analytic epistemology. This means that the conceptual content of epistemologized psychology should not be assumed *a priori* to have substantial overlap with the conceptual content of contemporary analytic epistemology. Likewise, epistemologized psychology need not, and probably should not, have among its methodological ends the goal of determining whether or not the intuitions that philosophers treat as evidence are widely shared among people who are not professional philosophers (Stich, 2018; Machery et al., 2015; Nado, 2016; Weinberg, 2015); though it is essentially experimental, epistemologized psychology is not *that* kind of experimental epistemology. But at the same time, it is not unreasonable to predict that there may be some areas of conceptual overlap between empirically adequate theories in epistemologized psychology and the conceptual content of certain popular theories in analytic epistemology— since, after all, some of the best of these theories represent efforts to *psychologize* analytic epistemology (Sosa, 2017; Goldman, 2002).

Conclusion

We believe, thus, that the hypothesis that the mind constructs a theory of evidence early in life represents an intriguing area for future research in epistemologized psychology. Indeed, we think that more systematic efforts to understand the conceptual resources and inferential structure of the mind's intuitive epistemology represent an even richer area of future research. The payoff, philosophically speaking, for carrying out either of these research programs is that the success of either would contribute to the further naturalization of epistemology by way of increasing the epistemologization of psychology. Indeed, it is not impossible that epistemologized psychology may eventually be able to answer questions such as "What epistemological concepts and principles do people have, in virtue of which they are able to learn?" "How does a theory of evidence develop over time?" "What elements of the mind's intuitive epistemology are most conducive to learning?" "What forms of rationality are possible for the human mind?" and so on.

The upshot, therefore, to further epistemologizing psychology is that we may eventually arrive at compelling answers to very deep questions about the mind, belief, and knowledge—and where these answers do not differ in their respective degrees of scientific and philosophical plausibility.

Notes

1 In fact, this is an oversimplification. Many of things that count as evidence cannot be classified correctly as the thing that they are based on observation alone. Instead, many of the things that a learner can learn to use as evidence can only be accurately classified using any number of independent psychological theories or abstract concepts. For example, word learning depends a theory of language, or at least a minimally useful semantic theory. So, a more precise formulation of our hypothesis is that a separate theory of evidence is needed to explain what other domain-specific psychological theories cannot—viz., children's ability to identify what things are appropriately classified as evidence but nevertheless fall within the scope of the other theories that they know. The point here in this footnote is to clarify that sometimes—perhaps all the time—this may involve abstracting over an abstraction.

2 Thus, one of the crucial functions that a child's theory of evidence is that it allows her to determine what information is and is not epistemically relevant to her learning. This is not an easy problem to solve. But it is an easier problem than a related and, to epistemologists, much more familiar problem. Recent work in philosophical epistemology has examined how the ability to determine if a piece of information (a fact, a proposition, etc.) is or is not epistemically relevant can help answer the Cartesian sceptic and thus underwrite a very strong conception of knowledge. Indeed, some well-known examples in the literature hold that learning—if learning produces knowledge—requires a knower be able to eliminate *all* relevant alternatives, and so requires of learners the underlying psychological ability to determine all possible relevant criteria. For example, David Lewis proposed that seven independent principles can be used to determine whether information is epistemically relevant or irrelevant—and being able to use these criteria involve, inter alia, making judgments about the reliability of abductive methods and the scope of established conventions for ignoring past knowledge (Lewis, 1996). Perhaps using Lewis' principles are some of what is required to construct a theory of knowledge that can resist a Cartesian skeptical. But as a matter of psychological plausibility, we think it is much too demanding to impute the capacity to use most

Lewis' principles to young children. So, the epistemologist's problem of relevance is not the same problem of relevance that we believe a theory of evidence purports to solve. Instead, our problem is to explain how the learning process starts, which, again, is not to be confused with the problem of constructing a form of knowledge that is immune to Cartesian skepticism.

References

Begus, K. and Southgate, V. (2012). Infant pointing serves an interrogative function. *Developmental Science*, 15(5): 611–17.

Behne, T., Liszkowski, U., Carpenter, M., and Tomasello, M. . (2012). Twelve-month-olds' comprehension and production of pointing. *The British Journal of Developmental Psychology*, 30(3): 359–75.

Birch, S. A. J., Vauthier, S. A. and Bloom, P. (2008). Three- and four-year-olds spontaneously use others' past performance to guide their learning. *Cognition*, 107(3): 1018–34.

Bonawitz, E., Shafto, P., Gweon, H., Goodman, N. D., Spelke, E., and Schulz, L. l. (2011). The double-edged sword of pedagogy: Instruction limits spontaneous exploration and discovery. *Cognition*, 120(3): 322–30.

Brugger, A., Lariviere, L. A., Mumme, D. L., and Bushnell, E. W. (2007). Doing the right thing: Infants' selection of actions to imitate from observed event sequences. *Child Development*, 78(3): 806–24.

Buchsbaum, D., Gopnik, A., Griffiths, T. L., and Shafto, P. (2011). Children's imitation of causal action sequences is influenced by statistical and pedagogical evidence. *Cognition*, 120(3): 331–40.

Butler, L. P. and Markman, E. M. (2012). Preschoolers use intentional and pedagogical cues to guide inductive inferences and exploration. *Child Development*, 83(4): 1416–28.

Butler, L. P. and Markman, E. M. (2014). Preschoolers use pedagogical cues to guide radical reorganization of category knowledge. *Cognition*, 130(1): 116–27.

Corriveau, K. H., Fusaro, M., and Harris, P. L. (2009). Going with the flow: Preschoolers prefer nondissenters as informants. *Psychological Science*, 20(3): 372–77.

Csibra, G. and Gergely, G. (2009). Natural pedagogy. *Trends in Cognitive Sciences*, 13(4): 148–53.

Csibra, G. and Gergely, G. (2011). Natural pedagogy as evolutionary adaptation. *Philosophical Transactions of the Royal Society of London. Series B, Biological Sciences*, 366(1567): 1149–57.

Danovitch, J. H. and Keil, F. C. (2004). Should you ask a fisherman or a biologist?: Developmental shifts in ways of clustering knowledge. *Child Development*, 75(3): 918–31.

Denison, S. and Xu, F. (2010). Twelve- to 14-month-old infants can predict single-event probability with large set sizes. *Developmental Science*, 13(5): 798–803.

Denison, S. and Xu, F. (2012). Probabilistic inference in human infants. *Advances in Child Development and Behavior*, 43: 27–58.

Fedyk, M. and Xu, F. (2017). The epistemology of rational constructivism. *Review of Philosophy and Psychology*, Vol 9, Issue 2, 343-362 .

Feldman, R. (2011). Naturalized epistemology. *Stanford Encyclopedia of Philosophy*. Available at: https://stanford.library.sydney.edu.au/archives/win2011/entries/epistemology-naturalized/

Furrow, D., Moore, C., Davidge, J., and Chiasson, L. (1992). Mental terms in mothers' and children's speech: Similarities and relationships. *Journal of Child Language*, 19(3): 617–31.

Goldman, A. I., 2002. *Pathways to Knowledge: Private and Public*. Oxford: Oxford University Press.

Gopnik, A. and Meltzoff, A. N. (1997). *Words, Thoughts, and Theories*. Cambridge: The MIT Press.

Henderson, A. M. E. and Sabbagh, M. A. (2010). Parents' use of conventional and unconventional labels in conversations with their preschoolers. *Journal of Child Language*, 37(4): 793–816.

Hetherington, C., Hendrickson, C., and Koenig, M. (2014). Reducing an in-group bias in preschool children: The impact of moral behavior. *Developmental Science*, 17(6): 1042–49.

Jaswal, V. K. and Neely, L. A. (2006). Adults don't always know best: Preschoolers use past reliability over age when learning new words. *Psychological Science*, 17(9): 757–58.

Johnsen, B. C. (2005). How to read "Epistemology Naturalized". *The Journal of Philosophy*, 102(2): 78–93.

Johnson-Laird, P. N. (1988). *The Computer and the Mind: An Introduction to Cognitive Science*. Cambridge: Harvard University Press.

Kinzler, K. D., Corriveau, K. H., and Harris, P. L. (2011). Children's selective trust in native-accented speakers. *Developmental Science*, 14(1): 106–11.

Koenig, M. A., Clément, F., and Harris, P. L. (2004). Trust in testimony: Children's use of true and false statements. *Psychological Science*, 15(10): 694–98.

Koenig, M. A. and Harris, P. L. (2005). Preschoolers mistrust ignorant and inaccurate speakers. *Child Development*, 76(6): 1261–77.

Koenig, M. A. and Harris, P. L. (2007). The basis of epistemic trust: Reliable testimony or reliable sources? *Episteme; rivista critica di storia delle scienze mediche e biologiche*, 4(3): 264–84.

Kornblith, H. (2002). *Knowledge and Its Place in Nature*. Oxford: Clarendon Press.

Kushnir, T., Vredenburgh, C., and Schneider, L. A. (2013). "Who can help me fix this toy?" The distinction between causal knowledge and word knowledge guides preschoolers' selective requests for information. *Developmental Psychology*, 49(3): 446.

Kushnir, T., Wellman, H. M., and Gelman, S. A. (2008). The role of preschoolers' social understanding in evaluating the informativeness of causal interventions. *Cognition*, 107(3): 1084–92.

Lackey, J. (2008). *Learning from Words: Testimony as a Source of Knowledge*. Oxford: Oxford University Press.

Lewis, D. (1996). Elusive knowledge. *Australasian Journal of Philosophy*, 74(4): 549–67.

Liszkowski, U., Carpenter, M., and Tomasello, M. (2008). Twelve-month-olds communicate helpfully and appropriately for knowledgeable and ignorant partners. *Cognition*, 108(3): 732–39.

Longino, H. E. (1996). Cognitive and non-cognitive values in science: Rethinking the dichotomy. In L. H. Nelson and J. Nelson (Eds.), *Feminism, Science, and the Philosophy of Science* (pp. 39–58). Synthese Library. Springer Netherlands.

Lutz, D. J. and Keil, F. C. (2002). Early understanding of the division of cognitive labor. *Child Development*, 73(4): 1073–84.

Machery, E. et al. (2015). Gettier across cultures. *Noûs*. Available at: http://onlinelibrary.wiley.com/doi/10.1111/nous.12110/full

Marr, D. (1982). *Vision*. Cambridge: MIT Press.

Mascaro, O. and Sperber, D. (2009). The moral, epistemic, and mindreading components of children's vigilance towards deception. *Cognition*, 112(3): 367–80.

Mills, C. M. et al. (2011). Determining who to question, what to ask, and how much information to ask for: The development of inquiry in young children. *Journal of Experimental Child Psychology*, 110(4): 539–60.

Moll, H. and Tomasello, M. (2007). Cooperation and human cognition: The Vygotskian intelligence hypothesis. *Philosophical Transactions of the Royal Society of London. Series B, Biological sciences*, 362(1480): 639–48.

Nado, J. (2016). The intuition deniers. *Philosophical Studies*, 173(3): 781–800.

O'Neill, D. K. (1996). Two-year-old children's sensitivity to a parent's knowledge state when making requests. *Child Development*, 67(2): 659–77.

Pasquini, E. S., Corriveau, K. H., Koenig, M., and Harris, P. L. (2007). Preschoolers monitor the relative accuracy of informants. *Developmental Psychology*, 43(5): 1216–26.

Putnam, H. (1982). Why reason can't be naturalized. *Synthese*, 52(1): 3–23.

Quine, W. V. O. (1969). Epistemology naturalized. In W. V. O. Quine (Ed.), *Ontological Relativity and Other Essays*. New York: Columbia University Press.

Sabbagh, M. A. and Baldwin, D. A. (2001). Learning words from knowledgeable versus ignorant speakers: Links between preschoolers' theory of mind and semantic development. *Child Development*, 72(4): 1054–70.

Sabbagh, M. A. and Shafman, D. (2009). How children block learning from ignorant speakers. *Cognition*, 112(3): 415–22.

Shatz, M., Wellman, H. M., and Silber, S. (1983). The acquisition of mental verbs: A systematic investigation of the first reference to mental state. *Cognition*, 14(3): 301–21.

Sobel, D. M. and Corriveau, K. H. (2010). Children monitor individuals' expertise for word learning. *Child Development*, 81(2): 669–79.

Sosa, E. (2017). *Judgment and Agency*, 1 ed. Oxford: Oxford University Press.

Southgate, V., Chevallier, C., and Csibra, G. (2009). Sensitivity to communicative relevance tells young children what to imitate. *Developmental Science*, 12(6): 1013–19.

Southgate, V., van Maanen, C., and Csibra, G. (2007). Infant pointing: Communication to cooperate or communication to learn? *Child Development*, 78(3): 735–40.

Stephens, E. C. and Koenig, M. A. (2015). Varieties of testimony: Children's selective learning in semantic versus episodic domains. *Cognition*, 137: 182–88.

Stich, S. (2018). Knowledge, intuition, and culture. In J. Proust and M. Fortier (Eds.), *Metacognitive Diversity: An Interdisciplinary Approach*. Oxford: Oxford University Press.

Tenenbaum, J. B. et al. (2011). How to grow a mind: Statistics, structure, and abstraction. *Science*, 331(6022): 1279–85.

Tomasello, M. (2014). *A Natural History of Human Thinking*. Cambridge: Harvard University Press.

Weinberg, J. M. (2015). Humans as instruments: Or, the inevitability of experimental philosophy. In *Experimental Philosophy, Rationalism, and Naturalism* (pp. 179–195). New York: Routledge.

Williamson, R. A., Jaswal, V. K., and Meltzoff, A. N. (2010). Learning the rules: Observation and imitation of a sorting strategy by 36-month-old children. *Developmental Psychology*, 46(1): 57–65.

Xu, F. (2011). Rational constructivism, statistical inference, and core cognition. *The Behavioral and Brain Sciences*, 34(03): 151–52.

Xu, F. (2007). Rational statistical inference and cognitive development. *The Innate Mind: Foundations and the Future*, 3: 199–215.

Xu, F. and Denison, S. (2009). Statistical inference and sensitivity to sampling in 11-month-old infants. *Cognition*, 112(1): 97–104.

Xu, F. and Kushnir, T. (2013). Infants are rational constructivist learners. *Current Directions in Psychological Science*, 22(1): 28–32.

Xu, F. and Tenenbaum, J. B. (2007). Word learning as Bayesian inference. *Psychological Review*, 114(2): 245–72.

Yu, Y. and Kushnir, T. (2015). Understanding young children's imitative behavior from an individual differences perspective. *CogSci*. Available at: https://mindmodeling.org/cogsci2015/papers/0474/paper0474.pdf

Part Three

Special Sciences

7

Applying Experimental Philosophy to Investigate Economic Concepts: Choice, Preference, and Nudge

Michiru Nagatsu

Introduction

Philosophers of science discuss not only general epistemological and metaphysical questions about explanation, causation, evidence, and the like, but also conceptual questions concerning the nature of scientific concepts such as genes, culture, and rationality. One might expect less disagreements in the latter debates, since the philosophical analyses are presumably based on the same "best scientific theories" available at the moment. The disagreements over the exact nature of these scientific concepts, however, seem sometimes more fundamental than those over, for example, what constitute a good explanation. One might argue that this is healthy because a naturalistic philosophy of science should reflect genuine disputes in scientific practice. But such dissonance among scientists may be exacerbated by philosophical commentaries, because philosophers of science typically rely on different evidence bases in an unsystematic way: some consult their intuitions, others firsthand experience as practitioners in the relevant scientific discipline, and yet others a small number of case studies of research articles. Although a narrow focus on a particular type of evidence can deepen our understanding of some aspects of scientific practice, if uncoordinated, it fails to provide a big picture of the scientific conceptual landscape (Weinberg and Crowley, 2009). Even worse, it may provide a distorted image of science.

Experimental philosophy (X-phi) of science is a relatively new approach that aims to overcome this problem. Specifically, it uses survey-experimental instruments to generate data about scientists' judgments on conceptual issues, in a hypotheses-oriented and controlled fashion, thereby complementing or confronting the kinds of evidence mentioned here regarding how scientists understand and use particular concepts (Griffiths and Stotz, 2008; Machery, 2016). Until very recently, experimental philosophy has not been much applied in the philosophy of science, despite its popularity in other fields of philosophy.[1] X-phi is even less popular in philosophy of economics, compared to other scientific fields (e.g., Stotz, 2009; Linquist et al., 2011; Knobe and Samuels, 2013). I think this is just a contingent fact, and there is no deep reason that prevents an experimental approach from being useful in the philosophy of economics. I argued elsewhere that this unfortunate situation should change (Nagatsu, 2013) and conducted one of the first consciously X-phi of economics studies (Nagatsu and Põder, 2019). Drawing on these results, I will argue that an X-phi approach can indeed bring conceptual clarity to some debates in philosophy of economics.

Another related field is worth a brief mention before discussing X-phi of economics. Philosophical questions concerning economics include not only methodological and conceptual ones about economics as a science, but also theoretical and normative ones within economics, such as the nature of justice, welfare, norms, and conventions. This field overlaps with philosophy both in content and in style, being highly theoretical, abstract, sometimes formal, and with little empirical input other than theorists' intuitions. The rise of experimental and behavioral economics, however, changed this situation, just like X-phi has changed philosophy, and it has become increasingly popular to study these issues using experimental games of bargaining, coordination, and social dilemmas. I call this field *experimental economics of philosophy* to distinguish it from X-phi of economics, and to highlight the fact that its method comes from experimental economics, while the subject matter is of philosophical interest. Although it is in practice difficult to draw a distinct line between this and the rest of experimental economics, one can identify several studies with explicitly philosophical focus on, for example, justice (Konow, 2003), Humean and Lewisian conventions (Mehta, Starmer, and Sugden, 1994; Guala and Mittone, 2010; Guala, 2013), and moral judgments and behavior (Gold, Colman, and Pulford, 2014, 2015). Of particular interest for X-philes is the use of monetary incentives, one of the methodological features of experimental economics of philosophy that may be useful in other X-phi studies as well (Gold, Pulford, and Colman, 2013).

The chapter is organized as follows: first, I will introduce two working hypotheses concerning the variance and validity of scientific concepts, conceptual variance, and conceptual ecology hypotheses, which I borrow from X-phi of biology and draw on throughout the chapter ("Conceptual variance and conceptual ecology"). Then, I introduce commonsensible realism as the received view in philosophy of economics ("Commonsensible realism"). In "Choice concepts: folk vs. economic," I discuss folk vs. economic concepts of choice, drawing on my own study. In "Preference concepts: Behavioral, psychological, or constructive?" I discuss behavioral vs. mental interpretations of preferences that divide economists and psychologists. The penultimate section ("Tracking changing methodological practice: To nudge or not to nudge?") discusses how X-phi can shed light on conceptual ecology in addition to conceptual variance. A brief conclusion follows ("Conclusion").

Conceptual variance and conceptual ecology

The early studies in X-phi of science (in particular Griffiths and Stotz, 2008) provide a useful framework for the empirical-conceptual investigation of economic science. This framework consists of two working hypotheses: conceptual variance and conceptual ecology. The conceptual variance hypothesis states that a given concept may have different meanings across different scientific communities; the conceptual ecology hypothesis states that there are often methodological reasons, both epistemic and practical, for such variance. More generally, scientists adapt cognitive resources, such as models, concepts, and other techniques to their own specific problem-solving domains to facilitate their cognitive and practical goals. This adaptation, or *epistemic niche construction* (Sterelny, 2010), gives rise to domain specificity (MacLeod, 2018) of scientific practices. Conceptual variance can be understood as a manifestation of this domain specificity in scientific concepts, reflecting conceptual ecology.

In general, three types of conceptual variance can be distinguished: (i) folk vs. scientific variance, (ii) interdisciplinary variance, and (iii) intradisciplinary variance. The first concerns variance across lay and expert concepts, while the latter two concern variance across scientific communities, large and small, respectively. "Commonsensible realism" and "Choice concepts: Folk vs. economic" discuss (i); "Preference concepts: Behavioral, psychological, or constructive?" discusses (ii); and "Tracking changing methodological practice: To nudge or not to nudge?" discusses (iii).

Commonsensible realism

Traditionally, philosophers of economics have discussed theoretical concepts in economics (mostly in the theory of consumer choice) in the context of realism. Uskali Mäki has been one of the proponents of *commonsensible realism*, according to which the entities and relationships of economic theory are part of the "common-sense furniture of the human world" (Mäki, 2002b, p. 95). In particular, the class of psychological states posited by the theory of consumer choice is part of "the ontic furniture of common-sense psychology, which we all employ in our daily lives regardless of whether we have an academic degree in psychology" (Mäki, 2000, p. 111). Daniel Hausman concurs: the unobservables in economics, that is, "beliefs, preferences, and the like are venerable. They have been a part of common sense understanding of the world for millennia ... there is no principled epistemological divide between the beliefs and desires [of] everyday life and the subjective probabilities and utilities of economics" (Hausman, 1998, pp. 197–99).

Although the motivations of Mäki and Hausman are different, they both dismiss the relevance of the observable/unobservable distinction to the philosophy of economics; unlike in the philosophy of physics, they insist, this distinction is unimportant for scientific realism debates in economics. Mäki wants to shift philosophers' attention away from ontological questions to more specific, methodological questions regarding the representational strategies of economists, such as isolation, abstraction, and idealization, and how to evaluate these strategies necessarily involving unrealistic assumptions. The key question concerns the truth, or "realisticness" of representations, not the existence of postulated entities. Although Hausman argues that scientific realism, including the kind of realism Mäki sees as central, is largely irrelevant to economic methodology, both agree that commonsensible realism provides necessary and sufficient ontology for economic theory.

Contrary to this received view, I argue that commonsensible realism is a necessary starting point of economic methodology, but it is not sufficient. First of all, there is a relevant analogy between realism in physics and economics; although unobservable entities like electrons do not figure in economics, the exact nature of latent constructs have always been controversial. For example, the nature of preferences has been extensively debated in the philosophy of economics (see "Preference concepts: Behavioral, psychological, or constructive?"). Preferences or utility are said to be latent not because they are too small to be seen with the naked eye but because they have to be inferred from observable behavior. Despite this difference, their nontransparent character gives rise to disagreement concerning the conceptual and ontological nature of preferences, just like esoteric concepts and entities in physics do. Commonsensible realism, on a strong reading, cannot explain why researchers disagree on the nature of preferences—we wouldn't expect such a disagreement if economic constructs are identical with concepts which everyone shares from everyday experience. Moreover, commonsensible realism lacks a resource to explain patterns of the disagreement—if there is any—about the nature of particular economic constructs. I show some evidence against the strong reading of commonsensible realism, and propose its weaker version as an alternative conceptual variance hypothesis in "Choice concepts: Folk vs. economic." Commonsensible realism needs to be qualified by accepting economic concepts' systematic departures from folk counterparts.

It is worth emphasizing at this stage that X-phi of economics is distinguished from general surveys on the opinions of economists and other members of the general public. A well-known study demonstrated the systematic gap between folk and economists' opinions about the economy. This is a series of telephone surveys of 1,511 noneconomists and 250 economists conducted by the Washington Post/Henry J. Kaiser Family Foundation/Harvard University Survey Project (Blendon et al., 1997). The main findings of this study include a systematic gap between economists' and lay people's reasoning about how the economy works. For example, lay people tend to see increased prices as a result of companies' price manipulation, while economists tend to see it as due to supply and demand. This and other findings reveal a systematic divergence between folk and economic theories of how the economy operates. But studies of this type cannot directly inform debates on commonsensible realism, because all the evidence shows is that lay people have their own folk-economic theory, which may be underscored by the common-sense concepts shared by the folk and economists. In order to investigate possible conceptual variance between folk and economists, a study needs to be designed to focus on concepts, not just opinions.

Choice concepts: Folk vs. economic

Ross (2011, p. 220) raises a thought-provoking challenge to commonsensible realism, speculating that those economists who are psychology or neuroscience skeptics "have a different concept of choice in mind" than that shared by noneconomists. This is surprising, because choice is not even latent or unobservable in the sense that preferences or beliefs are. Moreover, unlike these concepts that are formally defined in choice theory, the concept of choice itself is rarely explicitly characterized in the textbooks, as if people shared a common understanding of its meaning.[2] But if Ross is right, that is, if these economists have a concept of choice different from the one held by noneconomists, commonsensible realism needs to be qualified. This has practical implications, too. Regarding the gap between the folk and economic theories of the economy mentioned above, for instance, the gap might be more difficult to bridge than initially thought if the two groups diverge even at the supposedly commonsensical conceptual level.

Specifically, Ross states that choice is a pattern of behavior that varies in response to incentives. We highlight two points: first, on this view, choice in economics is a population-scale phenomenon, rather than an individual psychological one. This implies that choice can have very heterogeneous causal bases if one zooms in to the individual scale, such as effortful decision-making, imitation, inertia, constrained random behavior, and so on. Second, since these causal bases are heterogeneous, the subject may or may not be conscious that she is making a choice. Consciousness is thus not a necessary (nor a sufficient) condition for behavior to count as choice in this economic sense. Nevertheless, choice has to vary in response to shifts in incentives in a theoretically tractable way.

Ross's conceptual variance hypotheses can be reformulated as follows:

- H1: Economists are more likely than noneconomists to think of a behavioral change as choice, if it is a response to incentive shifts.
- H2: Economists are more likely than noneconomists to think of a behavioral change as choice, even if the actor is not aware that she is responding to incentive shifts.

In order to test these hypotheses by eliciting the respondents' notions of choice, we constructed two sets of *vignettes*, that is, stylized descriptions of hypothetical scenarios in which the protagonist changes his or her behavior prompted by a range of events (Nagatsu and Põder, 2019). We tested the two hypotheses in two different sets of vignettes. In the first set, we manipulated the dimension concerning the cause of the protagonist's behavioral change (Linda's reduce meat consumption). We had four levels, namely, (i) belief change, (ii) price change, (iii) medical change, and (iv) "nudged" change. In the second set, we manipulated the dimension concerning the protagonist's awareness of the cause of their own behavioral change (John stops winking to his female colleagues), where the cause is fixed as an incentive change (frowns of disapproval by winkees). We had three levels, namely, (i) being aware of the cause, (ii) being unaware of it, and (iii) interrupted by a cause overdetermining behavior regardless of awareness.

The order of the two sets were fixed, but the order of the vignettes within each set was randomized for different subjects.

We constructed two separate sets of vignettes (seven in total) instead of twelve by manipulating the two dimensions in 4 × 3 factorial design.[3] The main reason for this choice is that some levels in the two dimensions are not independent and create implausible vignette cases.[4] Thus we tested two dimensions, one at a time, in two separate sets of vignettes: the first set investigated the connection between the notion of choice and types of cause of behavioral change; the second set of vignettes investigated the connection between the notion of choice and what mediates behavioral change. At the end of each vignette, subjects were asked to agree or disagree to the statement: "Linda chose to eat less meat" and "John chose to stop winking," respectively, on a 7-point Likert scale.

We disseminated the online survey using Qualtrics (www.qualtrics.com). The link to the survey was disseminated using mailing lists at different universities in five countries: the United Kingdom (University of Reading), Finland (University of Helsinki, Hanken School of Economics), Estonia (Tallinn University of Technology), Italy (University of Milan), and Turkey (Bahcesehir University). The survey was also sent to the students who were enrolled in the course Understanding Economic Models (Fall semester 2016) at the Department of Political and Economic Studies, the University of Helsinki, before the course had started. Of the 185 respondents who started, 127 completed the survey (completion rate was 69 percent; mean time for completion was 8 minutes). We did not give incentives in money or course grade. The main part of the survey was followed by demographic questions, including the main area of study, the level of education (BA, MA, and PhD), mother tongue, and gender, as well as a prompt to leave any comments on the survey in free form. The characteristics of the respondents are summarized in Table 7.1. We operationalized "economists" as those who selected "Economics" as the main area of study. "Business and Management" is distinguished from "Economics."

To summarize, we have two predictions:

- Prediction 1: other things being equal, economists (defined by their Main Area of Study) are more likely than noneconomists to agree to the statement "Linda chose to eat less meat" in the price change scenario but not in the others.
- Prediction 2: other things being equal, economists (defined by their Main Area of Study) are more likely than noneconomists to agree to the statement "John chose to stop winking" in the unconscious scenario but not in the others.

Table 7.1 Respondents' characteristics

Area of study		Education		Language		Gender	
Economics	73	BA	34	English	16	Male	79
Others	54	MA	25	Estonian	16	Female	48
		PhD	63	Finnish	34		
		n/a	5	Italian	23		
				Turkish	17		
				Others	21		
Total	127		127		127		127

Our regression analysis is largely consistent with our predictions and support Ross's hypotheses. In particular, answering "Economics" as the main area of study makes one more likely to judge Linda's incentive-induced behavioral change as a choice. This effect, call it the *economist effect*, is large (about 20 percent average marginal effects), statistically significant ($p < 0.05$), and robust (the effect size remains the same regardless of the exact thresholds for responses to be categorized as positive or negative answer), confirming Prediction 1. The results strongly suggest that economists are more likely than noneconomists to think of a behavioral change as choice if it is a response to incentive shifts. This supports Ross's hypothesis 1, as formulated above. The economist effect on judging John's unconscious incentive-induced behavioral change as choice is not as clear, and therefore we focus on the first economist effect here.[5]

Economists might have a distinctive, technical concept of choice that they apply in their scientific practice, but how do we know that the economist effect we observed reflect that methodologically relevant concept? The answer we advance is a specific version of commonsensible realism. Unlike the strong version, which identifies economic concepts with folk ones, this weak version accepts that economic concepts such as choice (and subjective beliefs, preferences, and the like) are continuous with common-sense counterparts but deviate from them in ways that reflect economic theoretical frameworks, such as the theory of choice. In other words, economists share some common-sense understanding of these concepts, which is overridden or partially modified by scientific disciplinary training (or alternatively purified by self-selective recruitment). While the strong version does not motivate empirical investigations of economic concepts (because we already know them from our everyday experience), the weak version motivates such investigations and also offers a plausible explanation of our observations that economists' concept of choice is affected by their theoretical frameworks.

The crucial questions are what constitutes the core "commonsensible" part of the choice concept, and what makes the economic concept of choice deviate from it. We hypothesized voluntariness as the core commonsensible of choice. Common sense tells economists (as well as noneconomists) that choice has to be voluntarily made—otherwise, you have no choice! In this sense, Linda's reduced meat consumption due to the increase in meat prices is less of a choice because of the limits imposed on a range of available options, thereby compromising the voluntary nature of her reduced meat consumption. According to the standard economic framework, however, choices simply reflect or "reveal" the subject's satisfaction of exogenous (i.e., given) preferences under certain constraints. In this framework, Linda's reduced meat consumption in response to price increase is a choice because her behavioral change still satisfies her preferences under a new, tighter budget constraint. This speculation provides a plausible mechanistic explanation of the economist effect we observed in the Linda vignette: while noneconomists interpreted the increase in prices as reducing the voluntary nature of Linda's response, economists did not, because their theory-laden concept of choice told them that it was irrelevant. Some might have explicitly thought: "Linda could have maintained the same level of meat consumption by, for example, buying less clothes." To sum up, this study suggests that commonsensible realism in the strong sense needs to be abandoned, while its weaker version is a plausible cognitive hypothesis regarding how and why economists' concept deviates from their folk counterpart. In the next section, I turn to the *preference* concept, whose nature has been disputed between economics and psychology.

Preference concepts: Behavioral, psychological, or constructive?

The notions of preferences and utility are among the most contested ones in the history and philosophy of economics. Historians of economic thought have discussed the development of these notions from the nineteenth to the twentieth century before World War II (Moscati, 2013; Lewin, 1996; Hands, 2012). More recently, the rise of behavioral economics and neuroeconomics revived this debate. While the discussion of commonsensible realism in "Choice concepts:

Folk vs. economic" concerns the variance between folk and scientific (economic) concepts, these debates on preferences explicitly concern variance of preference concepts in two scientific disciplines: economics and psychology. Roughly speaking, the debates take place in the context where psychology challenges economics for its lack of psychological realism. In this section, I will provide a brief overview of these debates and discuss how X-phi of economics can shed light on them.

The postwar development of revealed preference theory enabled economists to model choice behavior as utility maximization based on observable choice data and a set of parsimonious axioms regarding preference relations. Accordingly, the hedonic connotation of the utility concept—intensity of pleasures and pains—has been stripped away, and it has become simply a convenient way of indexing preferences that satisfy the axioms of revealed preference theory. There is a popular historical narrative among economists according to which this theoretical achievement is seen as a completion of the long-term separation of economics from psychological hedonism, under the influence of the contemporary behaviorism in the early twentieth century. Edwards (2016) calls it a "behaviorist myth"; however, this interpretation is misleading. First, strictly speaking, the notion of preferences never disappeared from choice theory, as originally envisioned by the young Paul Samuelson, the founder of revealed preference theory (Hands, 2010). Second, the core axioms of the theory (completeness and transitivity of preferences) are postulated *a priori*, not based on observations of human behavior using behaviorist methods such as conditioning. Third, Expected Utility Theory (Von Neumann and Morgenstern, 2004, first appeared in the second edition in 1947)—the standard theory of choice under risk and uncertainty—did not develop along the behaviorist line, either. This theory involves the notion of risk preference as a key construct, and Savage (1954) added an extra psychological construct of beliefs qua subjective probabilities. So, as a historical account, the behaviorist myth is just that, a myth. However, the myth may reflect an inherently ambivalent nature of the notion of preference in economics: on the one hand, the theory enables economists to infer (or *reveal*, as they say) preferences from observed choice data without data on mental or neurological processes. In this sense, the preference concept does not need psychology. On the other hand, the very axioms are based on *a priori* postulates about preference relations (and relations between subjective beliefs and preferences in the case of expected utility theory), which seem to be based on the introspective psychology of decision-making. Are preferences behavioral or psychological (mental)?

As noted earlier, the debates over this question have been intensified by the rise of behavioral economics in the last quarter of the twentieth century and neuroeconomics in the beginning of the twenty-first century. Behavioral economics, mostly influenced by cognitive psychology, has demonstrated numerous empirical anomalies to expected utility theory and game theory; some neuroeconomists even go further and suggest that the very notion of preferences may have to be abandoned given the data from new neuroimaging techniques such as fMRI (Camerer, Loewenstein, and Prelec, 2005). In response to the increasing pressure to revise mainstream economic theories based on these new types of data, Gul and Pesendorfer (2008) argued that psychological and neurophysiological evidence is simply irrelevant to the economic theory of choice because the theory does not refer to mental states and therefore is "mindless." Gul and Pesendorfer (2008) initiated the neuroeconomics controversy among economists and methodologists. (For a very careful and thorough methodological analysis, see the target article Harrison, 2008, as well as the commentaries in the same issue). Just like the behaviorist myth, the controversy is a symptom of a deep disagreement between psychological (mental) and behavioral interpretations of preferences.

To focus on conceptual issues, I turn to philosophers of economics who try to explicitly define what preferences are. Hausman (2012) provides one of the clearest conceptual analyses of preferences, according to which preferences in economics are total subjective comparative evaluations.[6] That is, they capture the agent's subjective rankings of available but competing alternatives after taking into account all relevant pro tanto reasons, which, jointly with her beliefs, cause (and justify and explain) her choice of one alternative over the other(s). So if Anne has two feasible evening plans, b (going to a friend's barbecue party) and c (going to the cinema), and prefers c to b, and if her beliefs do not interrupt with this preference (e.g., by reminding her that the cinema is closed today), she will and should choose c over b, because her preference (c ≻ b) has been formed by considering all the subjective factors that are relevant to her decision-making. Angner (2018) rejects Hausman's interpretation of preferences, based on the evidence that such mentalistic commitment on the part of economists cannot be found in (i) orthodox economics textbooks, (ii) commentaries by founding economists of the postwar neoclassical synthesis, or (iii) contemporary economists' practices.[7]

Angner's case against the mentalistic interpretation of the preference concept may seem to sway the balance between mental vs. behavioral interpretations toward the latter. In fact, however, Angner (2018) proposes the dissolution of the dichotomy by proposing a minimalist interpretation: preferences in economics are whatever the axioms of utility theory say they are. According to Angner, the preference concept is "implicitly defined" by virtue of its place in the axioms of choice theory and therefore has no intrinsic, definite meaning. This contextual view implies, first, that preference concept's definition can (implicitly) change as some axioms are modified; and second, even taking all the axioms as fixed, the empirical meaning of preferences may change depending on the domain to which the theory is applied by some correspondence rules.

Guala (2017) also denies the dichotomy between mental vs. behavioral interpretations of preferences and proposes a third interpretation, according to which preferences are dispositions with multiply realizable causal bases.[8] This view is not behavioral because preferences may have mental causal bases. For example, a consumer's intransitive preference over three cars with three attributes may be caused by her use of a simple heuristic for pairwise comparisons: choose one that beats the other on most attributes. However, the view is not mentalistic, either, because preferences may have nonmental causal bases. For example, a three-member committee's intransitive preference over three candidates (a choice problem formally equivalent to the last example) is caused by institutional rules: majority voting through a sequence of pairwise comparisons of the candidates. In both cases, preferences are dispositions that explain choice that have multiply realizable causal bases, such as mental rules of thumb and institutional rules of aggregation.

Guala's example of the intransitive preferences of the consumer and the committee can be seen as an elaboration on the second implication of Angner's contextual account of preferences; that is, the meaning of preferences depends on the domain of application of choice theory. More generally, Guala identifies three conditions for preference-based choice theories to be explanatory of an agent's behavior: the agent in question is (1) consequence-driven, (2) motivated to pursue different goals, and (3) able to compare the values of such goals. Note that these conditions themselves should not be given mentalistic interpretations. So predicates such as "is driven by," "is motivated to," and "is able to compare" are all applicable to nonhuman agents (neurons, bees, pigeons, etc.) and agents composed of human agents (committees, nation states, etc.), as well as to human agents with limited cognitive capacities (infants and boundedly rational people like us). In fact, conditions (2) and (3) come from the axioms concerning how preferences are ordered, for example, $b \succcurlyeq c$ or $c \preccurlyeq b$, and condition (1) comes from a particular correspondence rule used to infer preferences from observable behavior. We do not know *a priori* which agents' behavior fits the bill, and this seems to be the point at which philosophers have to hand the matter over to scientists.

Taken together, these two accounts of preferences in economics—Angner's Nagenlian contextual account and Guala's Fodorian multiple realizability account—offer a convincing argument to the effect that there is no definite answer to the question "are preferences mental (psychological) or behavioral?" There is nothing in the nature of preferences as such that can determine the answer, because it depends on how the axioms of choice theory implicitly define preference relations, and which domains the theory can be successfully applied to.

So if there is no philosophical disputes to be adjudicated here, what roles are left to X-philes of economics? I see at least two important roles of X-phi of economics here. First of all, the X-phi approach can show whether there is any systematic connection between the folk concept of preferences and its theory-laden, economic counterpart. Angner (2018, p. 21) suggests that everyday connotations of theoretical constructs are simply irrelevant to scientific debates, citing Nagel's position that "such connotations are irrelevant . . . and are best ignored" (Nagel, 1961, pp. 91–92). However, this claim will have to be qualified if the weak version of commonsensible realism I proposed in "Choice concepts: Folk vs. economic" also applies to preferences. The folk concept of preferences is relevant to the understanding of economic counterpart if the latter's departures from the former are systematic. And in order to know if such a systematic connection exists, we need to empirically investigate what the commonsensible of preferences is, for which the X-phi approach will be necessary. Moreover, the commonsensible of preferences is relevant because preference-based choice theory is simultaneously used as a basis of normative welfare evaluations of agents when the theory is applied to individual human agents. In this specific but main domain of applications, then, economists are imposing their implicit definition of preferences on individuals. And it is not clear why the individuals' own conception of preferences are "best ignored" in this normative context.

Second, X-phi can be useful in investigating methodologically relevant interdisciplinary differences between economics and psychology, which may persist even after the dichotomy between behaviorism and mentalism has been dissolved. To see this, let us restrict the domain of application of choice theory to individual humans again.

In this domain (including consumer choice theory), the causal basis of preferences is most naturally interpreted as mental, as suggested by Guala's (2017) example of heuristic-based intransitive preferences, and also Angner's

(2018) use of findings from cognitive science. Based on "far-reaching similarities" (Angner, 2018, p. 17) between how Hausman and cognitive scientists talk about preferences, Angner takes the findings of implicit biases (such as mere exposure effects) in cognitive science to be the evidence against Hausman's account of what preferences are in economics. This move is based on the assumption that what preferences are in this domain depends on what cognitive scientists say they are. But some economists disagree with even this localized mentalistic interpretation of preferences. Guala (2017) tries to dissolve this tension by suggesting that preferences are explanans for economists and explanandum for psychologists, that is, that economists explain individual choice based on preferences, which are in turn explained in terms of individual mental processes by psychologists. Ross (2014), however, questions such a neat division of explanatory labor. The economic concept of preferences, argues Ross, does not map onto individual-scale mental processes, which are too heterogenous to provide a generalization about preferences; nevertheless, statistical analysis of a large number of pooled individual choice data can reveal projectible preferences at a market scale. Such projectability comes largely from the market structure in which individuals are embedded, that is, artificial environmental scaffolds such as information-processing technology and engineered (or emergent) institutional rules. Importantly, the market structure is not bounded by the limits of individual rationality. Rather, the market structure can be designed (or evolve) precisely to transcend the limitations of individual decision-making rationality. The revealed rational patterns, such as systematic responses to incentives, are thus an ecological property of the market, not of any single individuals.

Note that Ross's account of preferences—which is analogous to his account of choice discussed in "Choice concepts: Folk vs. economic"—is not strictly behavioristic because it leaves room for the psychology of latent individual mental processes to play a causal role in market phenomena.[9] What he is resisting here is rather the prospect that psychology (or neuroscience for that matter) will eventually provide a general theory of what preferences are in economists' sense, supplanting the economic theory of choice.

If Ross (2011) is correct, in the domain of individual choice, economists are committed to the concept of preferences that are neither mental nor behavioral, but constructive: preferences emerge from the combination of individuals who process information internally and the market structures that scaffold their decisions and interactions. This hypothesis is testable by X-phi approach, and if supported by evidence, it may provide a novel explanation of the conservative nature of economic practices. For example, somewhat surprisingly, economics education remains more or less the same before and after the rise of behavioral economics, despite the prestige it has won over the last decades within the economics profession and its popularity outside academia. Core economic theories (e.g., price theory) have not been modified accordingly.[10] Psychologists see this practice as unscientific, and economics students find it baffling.

Indeed, understanding this conservativeness of economics has been one of the leitmotifs in philosophy of economics: (How) can we methodologically justify economists' reliance on psychologically unrealistic models of choice? Traditionally, this question has been addressed in the general epistemological frameworks such as idealization or ideal types (Angner, 2015), and isolation of target systems by models (see Mäki, 2002a, part III) or the robustness of such unrealistic models. These general frameworks, however, cannot make sense of the methodological uniqueness of economics vis-à-vis psychology, for they do not explain why economic idealization is different from the kind of idealization psychologists engage in.[11] In contrast, Ross's constructive characterization of preferences is based on his analysis of domain-specific methodology of economics, which general epistemological frameworks cannot penetrate. An X-phi study designed to identify economists' concept of preferences and its deviations from the folk and psychological counterparts may demonstrate such a domain specificity of the preference concept, which will motivate more fine-grained methodological discussions. For example, we can start from a survey-experimental test of Hausman's 2012 account of preferences in economics as total subjective comparative evaluations, which Angner (2018) criticizes. Each dimension of Hausman's conceptual analysis can be operationalized and examined in a relatively straightforward way. A next step will be to test Ross's account of constructive preferences, which will require vignettes that draw directly on real economic modeling practices and a larger sample of economists. Such studies may demonstrate the domain specificity of economic methodology in a concrete manner, which can help other scientists and economists to better understand their methodological differences. We also expect the results to be relevant to economics pedagogy and the public understanding of economics as a science.

Tracking changing methodological practice: To nudge or not to nudge?

X-phi is a useful approach to discovering systematic variances between folk and scientific concepts and between psychological and economic concepts. Given the diversity of economics, we should expect that conceptual variance exists even within economics. A literature survey by Cowen (2004), for example, identifies a few variations of the rationality concept in different fields of economics, such as consumer choice theory, macroeconomics, experimental economics, and game theory. In addition, two cases studies, one on Contemporary Revealed Preference Theory (Hands, 2012) and the other on behavioral economics (Guala, 2012), suggest rather different understandings of preferences within economics. The conceptual ecology thesis justifies such a cohabitation of different concepts within science, whether across or within disciplines. However, some philosophers may think that X-phi is good at taking a static snapshot of conceptual landscape but incapable of analyzing dynamic methodological changes, should they arise in the same field. I have two responses, one generic and one specific.

First, it is uncontroversial that different empirical methods have different advantages and limitations, and complement each other. Somewhat simplifying, qualitative case studies, including historical analysis, detailed analysis of published articles, interviews, and participatory observations, are good at studying scientific practices in depth; in contrast, X-phi and bibliometrics (systematic analysis of published documents) are good at providing breadth, or an overview of conceptual landscape and trends and networks, respectively. Of course, X-phi is distinct from the other methods in being able to manipulate vignettes in a systematic, hypothesis-oriented way. But these methods are all needed to fully understand scientific concepts. To extend the contextual theory of scientific concepts a bit further, it is not difficult to see that concepts are implicitly defined not only by theoretical postulates and the rules of correspondence but also by the entire web of scientific practices, which Chang (2012) calls a *system of practice*. In our case of preferences in economics, the relevant practices include institutional engineering of market rules, welfare analysis of such policy interventions, and the normative individualism that underpins such welfare analysis. Qualitative methods are best suited for studying these practices in depth.[12]

Second, however, X-phi may provide methodologically relevant information about dynamic changes in economic methodology and elucidate the nature of accompanying disputes, if not dissolve them. In this section, I will use nudge paternalism as an illustrative case.

Nudges are defined as policy interventions in choice architecture—contexts in which individual choice takes place—that do not significantly change people's incentives or beliefs but still reliably produce welfare enhancing behavioral change (Thaler and Sunstein, 2008). Famous examples include the default change of pension choice (from opt-in to opt-out), and the automatic retirement savings mechanism (Save More Tommorrow™). These nudges are reverse engineered based on various psychological effects found by behavioral economists, such as loss aversion, framing effects, the status quo bias, and the present bias. There have been a great deal of debates around nudges among philosophers, economists, legal scholars, and policy makers, in particular regarding nudges' potential threats to individual liberty and autonomy and their ethical permissibility as a policy instrument. Economists are polarized in these debates, some embracing nudges and others rejecting them altogether. But it is not clear why and on what grounds they disagree. In particular, is the skepticism based on some moral conviction about individual autonomy and freedom or more subtle methodological (epistemic and practical) reasons? The conceptual ecology thesis and X-phi can illuminate the nature of this debate from a methodological perspective.

First, consider Ross's conceptual ecology thesis. Ross (2011) justifies his hypothesized variance between the economic, constructive concept of preferences and its psychological, individualistic counterpart, discussed in "Preference concepts: Behavioral, psychological, or constructive?" in terms of the practical concerns of respective fields as follows: psychologists are interested in the process of individual valuation and motivation because that is the scale on which most psychological interventions take effect, whereas economists are interested in the population-scale responses to incentive changes because that is the scale on which most economic interventions such as subsidies and taxes operate. At a very general level, this seems to be a plausible and useful characterization of the different epistemic and practical concerns of the two disciplines. And this conceptual ecological thesis provides a methodological insight into the debates around nudges.

Notice that Ross's (2011) own definition of preference and choice excludes nudged behavioral change from economic choice because by definition nudges are not incentives. However, there are significant conceptual similarities between incentives and nudges. First, both aim at population-scale behavioral change by indirectly intervening on the environment—constraints and choice architecture, respectively—rather than by directly intervening on individuals' motivations or cognition. Second, both are quiet about the exact psychological processes underlying behavioral change. Incentives presuppose that choices will respond to them, but there is tentative evidence that economists do not consider this process to be necessarily conscious, as discussed in "Choice concepts: Folk vs. economic" (see footnote 5). One might think that nudges are different because behavioral economists study nonstandard preferences caused by psychological biases, but note that many of the psychological effects I have listed here are experimental effects rather than causal mechanisms (see Guala, 2017, p. 9). Although several plausible psychological mechanisms have been proposed and tested in the lab as explanations of these effects, nudges are proposed based on their reliable effects at a population scale, rather than on the exact mechanisms that generate them. Third, both incentives and nudges presuppose exogenous preferences, though in crucially different ways. Incentives affect budget constraints, taking for granted people's preferences as given. Similarly, nudges change choice architecture, taking for granted people's preferences that nudges allegedly help manifest by removing psychological biases.

Given these similarities, it is no surprise that some pragmatic economists (e.g., Chetty, 2015) embrace nudges as part of economists' toolbox on par with incentives in behavioral public policy. In fact, there is not much in Ross's ecological thesis that prevents economists' adoption of nudges, given the latter's characteristics as (i) population-scale, (ii) mechanism-neutral, (iii) preference-reserving interventions. What do economists think? An X-phi study can examine whether and why economists (and others) accept nudges as legitimate policy interventions by systematically manipulating these factors.

In the X-phi study discussed in "Choice concepts: Folk vs. economic," we found evidence that economists' intuition about nudged behavioral change is not different from that of noneconomists. Nagatsu and Põder (2019) asked participants whether Linda's reduced meat consumption, nudged by the new arrangement of cafeteria food displays (another classic example of nudges in Thaler and Sunstein [2008]), was a choice. The mean answer was least positive among the four scenarios (4.51 on 1 "Strongly disagree" to 7 "Strongly agree") and not significantly different from noneconomists' responses (4.07, p(T-test) = 0.12, p(Wilcoxon Rank Sum test) = 0.23). This case, however, is not informative because the scenario was explicitly holding Linda's preference for meat as fixed. A better scenario should incorporate (iii) by making it explicit that Linda has a preference to reduce meat, which a nudge helps to satisfy. This way, X-phi can investigate not only whether economists accept nudges but also why.

So far, philosophers of economics have discussed nudges as ethical problems based on their understanding of what preferences are (Hausman and Welch, 2010; Bovens, 2009). Incidentally, both Hausman and Bovens subscribe to the concept of preferences as subjective total evaluations of alternatives, which Angner (2018) rejects as irrelevant to economics. I do not necessarily think that such ethical critiques are irrelevant because (a weak version of) commonsensible realism provides a link between folk and economic concepts of preferences in the domain of individual choice. But if economists see nudges as a population-scale intervention tool, ethical critics will have to justify why their ethical standards, based on a philosophical model of practical reasoning, have to weigh in economics. Also, philosophers focusing on mechanisms of nudges (Heilmann, 2014; Nagatsu, 2015; Grüne-Yanoff, 2016) will have to justify why mechanisms are important and in what level of details, if economists themselves (even behavioral economists) do not find issues of psychological mechanisms central for their methodology. We might be barking up the wrong trees.

Note that I am not arguing that philosophers of economics should refrain from participating in controversial contemporary debates in economics. On the contrary, I think that philosophers' participation can serve many functions such as providing conceptual clarification and outsiders' criticism. At the same time, there is a real danger that we philosophers will obscure the real issue by smuggling in our own perspectives to the debates, such as practical reasoning and mechanistic philosophy of science, in this case. The X-phi approach works as an antidote because it (i) reminds us of the domain specificity of economics (conceptual ecology), (ii) forces us to formulate hypotheses regarding the construction of economic concepts and how they are related to folk and other scientific concepts (conceptual variance), and (iii) gives us a way to test these hypotheses, in combination with evidence from other qualitative and nonexperimental studies of economics practices.

Conclusion

In this chapter, I have outlined what experimental philosophy of economics is and what it can do. It is an survey-experimental approach that illuminates conceptual issues in economics, driven by the conceptual variance and conceptual ecology as the working hypotheses. I have introduced our own study that suggests that economists think of choice in a systematically different way than noneconomists, specifically as a response to incentive shifts. I have suggested that this result can be best understood as a manifestation of the weak version of commonsensible realism; that is, the economic concept of choice is linked to the common-sense counterpart but departs from it in a systematic way. I have also discussed the concept of preferences that are contested between economists and psychologists, introducing their behavioral, mental, and constructive interpretations in the literature. I suggested that these interpretations are testable and may provide a novel understanding of economic methodology revealing its domain specificity. Finally, I have discussed the recent nudge debates as a case in which economists themselves disagree on the proper methodology of economics. Building on Ross's conceptual ecology thesis, I have highlighted three similar aspects of incentives (the standard tool of economic intervention) and nudges (a new and controversial tool), namely, their population-scale, mechanism-neutral, preference-reserving characters. We can analyze disagreements among economists at a deeper conceptual level by eliciting economists' responses by systematically varying vignettes along these three dimensions. At this stage, these ideas are more like thought experiments, with the exception of the real study of choice that we have conducted. And even that study has not covered a large enough, representative sample of economists. To do so will require more forces, so I welcome readers to join me in the exploration of this promising conceptual *terra incognita*.

Notes

1 In *PhilPapers* experimental philosophy of science does not have its own category; instead related papers are scattered across categories such as Foundations of Experimental Philosophy, Misc. and Experimental Philosophy, Misc. See https://philpapers.org/browse/experimental-philosophy. The situation is similar on popular blogs and edited volumes on X-phi.
2 In economics, choice and behavior are often used interchangeably, or the term *choice behavior* is used.
3 Griffiths, Machery, and Linquist, (2009) for example, used $2 \times 2 \times 2 = 8$ vignettes. More generally, in social research it is common to use more dimensions with more levels, resulting in a vast number of vignettes. Such design typically requires more subjects and random assignment of these subjects into different subsets of vignettes because it is practically impossible to expose each individual to more than a certain number of vignettes due to fatigue effects. Factorial surveys with this design are often not driven by clear hypotheses, unlike our case (see Nock and Guterbock, 2010, for a review).
4 See Auspurg and Hinz (2014, pp. 40–42) for the problems of implausible and illogical vignettes and ways to address them. In our case, (i) belief change and (iii) nudged change are respectively associated with conscious and unconscious mental processing: the agent is usually conscious about her behavioral change when she does so because of a changed belief; in contrast, if one changes behavior because of a nudge, often the change is not transparent to the agent. Discussions on the ethical permissibility of nudge-based behavioral policies revolve around this worry; see Thaler and Sunstein (2008).

5 The effect is large (about 18 percent average marginal effects) and significant (p < 0.05) but not robust, since the effect disappears in a model that is more permissive in categorizing responses as positive (this model takes 5, 6, and 7 on the 7-Likert scale as the positive answer). We can say that the economist effect exists only under the assumption that it should be detected in terms of a stronger agreement to the statement. Also, models that take levels of education as categorical rather than linear do not show the economist effect. Thus the second hypothesis—that economists are more likely than noneconomists to think of a behavioral change as choice even if the actor is not aware that she is responding to incentive shifts—is supported only under restrictive model assumptions.

6 Although Hausman does not cite it, Bovens (1992) offered a similar definition of preferences 20 years before in discussing the rationality of practical reasoning. Hausman's original contribution is to apply this definition of preferences to economics.

7 Angner also refers to empirical findings from cognitive science that show that people actually do not form preferences by integrating partial evaluations, and evolutionary explanations of why this has been adaptive.

8 See also (Ross, 2011). As one can imagine from our discussion in Choice concepts: Folk vs. economic, Ross has a very similar definition of preferences in economics.

9 Ross (2014, p. 416) is, however, critical about the empirical success and theoretical importance of the most famous model of boundely rational choice, Cumulative Prospect Theory (Tversky and Kahneman, 1992). I cannot discuss this substantial scientific debate here.

10 Some textbooks now have a chapter on behavioral economics, but it is treated more like an added topic, rather than a foundation of a complete revision of price theory.

11 Kahneman (2011) clearly indicates that psychological models such as the dual-system models are idealization.

12 See Wagenknecht, Nersessian, and Andersen (2015) for a general discussion of qualitative approach in philosophy of science. MacLeod and Nagatsu (2016, 2018) demonstrate the value of such methods in studying conditions for successful interdisciplinary model-building involving economists and ecologists.

References

Angner, E. (2015). To navigate safely in the vast sea of empirical facts. *Synthese*, 192(11): 3557–75.

Angner, E. (2018). What preferences really are. *Philosophy of Science*, 85: 660–81.

Auspurg, K. and Hinz, T. (2014). *Factorial Survey Experiments*. Los Angeles: Sage Publications.

Blendon, R. J., Benson, J. M., Brodie, M., Morin, R., Altman, D. E., Gitterman, D., Brossard, M., and James, M. (1997). Bridging the gap between the public's and economists' views of the economy. *The Journal of Economic Perspectives*, 11(3): 105–18.

Bovens, L. (1992). Sour grapes and character planning. *The Journal of Philosophy*, 89(2): 57–78.

Bovens, L. (2009). The ethics of nudge. In T. Grüne-Yanoff and S. O. Hansson (Eds.), *Preference Change: Approaches from Philosophy, Economics and Psychology* (Chapter 10, pp. 207–19). Berlin and New York: Springer.

Camerer, C., Loewenstein, G., and Prelec, D. (2005). Neuroeconomics: How neuroscience can inform economics. *Journal of Economic Literature*, 43(1): 9–64.

Chang, H. (2012). *Is Water H2O? Evidence, Pluralism and Realism. Boston Studies in the Philosophy of Science*. Springer: Dordrecht.

Chetty, R. (2015). Behavioral economics and public policy: A pragmatic perspective. *American Economic Review*, 105(5): 1–33.

Cowen, T. (2004). How do economists think about rationality? In M. Byron (Ed.), *Satisficing and Maximizing: Moral Theorists on Practical Reason* (Chapter 11, pp. 213–36). Cambridge: Cambridge University Press.

Edwards, J. (2016). Behaviorism and control in the history of economics and psychology. *History of Political Economy*, 48(suppl 1): 170–97.

Gold, N., Colman, A. M., and Pulford, B. D. (2014). Cultural differences in responses to real-life and hypothetical trolley problems. *Judgment and Decision Making*, 9(1): 65.

Gold, N., Pulford, B. D., and Colman, A. M. (2013). Your money or your life: Comparing judgements in trolley problems involving economic and emotional harms, injury and death. *Economics and Philosophy*, 29(2): 213–33.

Gold, N., Pulford, B. D., and Colman, A. M. (2015). Do as I say, don't do as I do: Differences in moral judgments do not translate into differences in decisions in real-life trolley problems. *Journal of Economic Psychology*, 47(Supplement C): 50–61.

Griffiths, P. E., Machery, E., and Linquist, S. (2009). The vernacular concept of innateness. *Mind and Language*, 24(5): 605–30.

Griffiths, P. E. and Stotz, K. (2008). Experimental philosophy of science. *Philosophy Compass*, 3(3): 507–21.

Grüne-Yanoff, T. (2016). Why behavioural policy needs mechanistic evidence. *Economics and Philosophy*, 32(3): 463–83.

Guala, F. (2012). Are preferences for real? Choice theory, folk psychology, and the hard case for commonsensible realism. In J. Kuorikoski, A. Lehtinen, and P. Ylikoski (Eds.), *Economics for Real: Uskali Mäki and the Place of Truth in Economic* (pp. 151–69). Abingdon-on-Thames: Routledge.

Guala, F. (2013). The normativity of Lewis conventions. *Synthese*, 190(15): 3107–22.

Guala, F. (2017). Preferences: Neither behavioural nor mental. DEMM Working Paper, Number 5.

Guala, F. and Mittone, L. (2010). How history and convention create norms: An experimental study. *Journal of Economic Psychology*, 31(4): 749–56.

Gul, F. and Pesendorfer, W. (2008). The case for mindless economics. In A. Caplin and A. Schotter, (Eds.), *The Foundations of Positive and Normative Economics: A Handbook* (chapter 1, pp. 3–42). Oxford: Oxford University Press.

Hands, D. W. (2010). Economics, psychology and the history of consumer choice theory. *Cambridge Journal of Economics*, 34(4): 633–48.

Hands, W. (2012). Realism, commonsensibles, and economics: The case of contemporary revealed preference theory. In P. Ylikoski, A. Lehtinen, and J. Kuorikoski (Eds.), *Economics for Real: Uskali Mäki and the Place of Truth in Economic* (pp. 156–178). Routledge.

Harrison, G. W. (2008). Neuroeconomics: A critical reconsideration. *Economics and Philosophy*, 24: 303–44.

Hausman, D. M. (1998). Problems with realism in economics. *Economics and Philosophy*, 14: 185–213.

Hausman, D. M. (2012). *Preference, Value, Choice, and Welfare*. Cambridge: Cambridge University Press.

Hausman, D. M. and Welch, B. (2010). Debate: To nudge or not to nudge. *The Journal of Political Philosophy*, 18(1): 123–36.

Heilmann, C. (2014). Success conditions for nudges: A methodological critique of libertarian paternalism. *European Journal for Philosophy of Science*, 4: 75–94.

Kahneman, D. (2011). *Thinking, Fast and Slow*. New York: Farrar, Straus and Giroux.

Knobe, J. and Samuels, R. (2013). Thinking like a scientist: Innateness as a case study. *Cognition*, 126(1): 72–86.

Konow, J. (2003). Which is the fairest one of all? A positive analysis of justice theories. *Journal of Economic Literature*, 41(4): 1188–1239.

Lewin, S. B. (1996). Economics and psychology: Lessons for our own day from the early twentieth century. *Journal of Economic Literature*, 34(3): 1293–323.

Linquist, S., Machery, E., Griffiths, P. E., and Stotz, K. (2011). Exploring the folkbiological concept of human nature. *Philosophical Transactions of the Royal Society: B Biological Sciences*, 366: 444–53.

Machery, E. (2016). Experimental philosophy of science. In W. Buckwalter and J. Sytsma (Eds.), *A Companion to Experimental Philosophy* (chapter 33, pp. 475–90). Chichester: Wiley-Blackwell.

MacLeod, M. (2018). What makes interdisciplinarity difficult? Some consequences of domain specificity in interdisciplinary practice. *Synthese*, 195(2): 697–720.

MacLeod, M. and Nagatsu, M. (2016). Model coupling in resource economics: Conditions for effective interdisciplinary collaboration. *Philosophy of Science*, 83: 412–33.

MacLeod, M. and Nagatsu, M. (2018). What does interdisciplinarity look like in practice: Mapping interdisciplinarity and its limits in the environmental sciences. *Studies in History and Philosophy of Science Part A*, 67: 74–84.

Mäki, U. (2000). Reclaiming relevant realism. *Journal of Economic Methodology*, 7(1): 109–25.

Mäki, U. (2002a). *Fact and Fiction in Economics: Models, Realism and Social Construction*. Cambridge: Cambridge University Press.

Mäki, U. (2002b). Some nonreasons for nonrealism about economics. In U. Mäki (Ed.), *Fact and Fiction in Economics: Models, Realism and Social Construction* (Chapter 4, pp. 90–104). Cambridge: Cambridge University Press.

Mehta, J., Starmer, C., and Sugden, R. (1994). Focal points in pure coordination games: An experimental investigation. *Theory and Decision*, 36(2): 163–85.

Moscati, I. (2013). Were Jevons, Menger, and Walras really cardinalists? on the notion of measurement in utility theory, psychology, mathematics and other disciplines, ca. 1870–1910. *History of Political Economy*, 45(3): 373–414.

Nagatsu, M. (2013). Experimental philosophy of economics. *Economics and Philosophy*, 29(2): 263–76.

Nagatsu, M. (2015). Social nudges: Their mechanisms and justification. *Review of Philosophy and Psychology*, 6: 481–94.

Nagatsu, M. and Põder, K. (2019). What is the economic concept of choice? An experimental philosophy study. *Economics and Philosophy*. under review.

Nagel, Ernest (1961). *The Structure of Science: Problems in the Logic of Scientific Explanation*. London: Routledge & Kegan Paul.

Nock, S. L. and Guterbock, T. M. (2010). Survey experiments. In P. V. Marsden and J. D. Wright (Eds.), *Handbook of Survey Research* (Chapter 28, 2nd ed., pp. 837–64). Bingley: Emerald Publishing Group Limited.

Ross, D. (2011). Estranged parents and a schizophrenic child: Choice in economics, psychology and neuroeconomics. *Journal of Economic Methodology*, 18: 217–31.

Ross, D. (2014). *Philosophy of Economics*. New York: Palgrave Macmillan.

Savage, L. J. (1954). *The Foundations of Statistics*. New York: Wiley.

Sterelny, K. (2010). Minds: Extended or scaffolded? *Phenomenology and the Cognitive Sciences*, 9(4): 465–81.

Stotz, K. (2009). Philosophy in the trenches: From naturalized to experimental philosophy (of science) introduction. *Studies In History and Philosophy of Science*, 40: 225–26.

Thaler, R. H. and Sunstein, C. R. (2008). *Nudge: Improving Decisions about Health, Wealth, and Happiness*. New Haven: Yale University Press.

Tversky, A. and Kahneman, D. (1992). Advances in prospect theory: Cumulative representation of utility. *Journal of Risk and Uncertainty*, 5: 297–323.

Von Neumann, J. and Morgenstern, O. (2004). *Theory of Games and Economic Behavior* (60th anniversary edition). Princeton: Princeton University Press.

Wagenknecht, S., Nersessian, N. J., and Andersen, H. (Eds.) (2015). *Empirical Philosophy of Science: Introducing Qualitative Methods into Philosophy of Science*. New York: Springer.

Weinberg, J. M. and Crowley, S. (2009). The X-phi(les): Unusual insights into the nature of inquiry. *Studies In History and Philosophy of Science*, 40: 227–32.

8

Scientists' Concepts of Innateness: Evolution or Attraction?

Edouard Machery, Paul Griffiths, Stefan Linquist, and Karola Stotz

The concept of innateness is an important component of *folkbiology* (Medin and Atran, 1999), the body of beliefs that people spontaneously rely on to make sense of their biological environment (reproduction, growth, decay, death, etc.). The innateness concept leads people to distinguish two kinds of biological traits: those that are innate and those that are learned. Innate traits are expressions of an organism's nature: they develop spontaneously and reliably (except under "abnormal" or "unnatural" conditions); they are shared by all members of the species or its natural subclasses, like males or juveniles (if an individual lacks the trait, they are "deformed" or "abnormal"); and they are functional in that they contribute to the life of the organism. Acquired traits, by contrast, are imposed, as it were, from the outside, and they are found in some, but not all, conspecifics.

The idea that members of a species share an essential nature that explains the typical character of the species has been attributed by philosophers to the influence of Plato and Aristotle (Hull, 1965). However, it is likely that these philosophers, like early naturalists, were merely codifying the common sense of their local culture (Atran, 1990). This simple framework for thinking about biological development allowed early thinkers to make sense of basic observations such as the tendency for offspring to resemble parents or for members of the same species to exhibit similar behaviors. It had a practical, heuristic value in the domestication of animals and selection for preferred traits. It predicted, for example, that dogs could be selected for loyalty, but not for loyalty to a particular family: The former is innate, the latter acquired. It is therefore not surprising that this prescientific distinction is deeply entrenched in folkbiology.

However, the concept of innateness appears to be incompatible with our scientifically informed understanding of evolution, development, and heredity.

We now recognize that all traits develop as a result of the interaction between genetic and environmental factors. It is therefore misleading to ask, "is this trait innate or acquired?" The more appropriate question is "how do specific genetic and environmental factors interact in the development of this trait?"

Moreover, as biologists uncover details about how certain traits develop, a further problem with the folk concept has become apparent. According to this concept, innate traits are expected to have three basic properties: developmental fixity, species typicality, and biological function. Thus, the presence of one property, on the folk view, is taken as evidence for the presence of the other two. For example, if a trait is developmentally buffered against some environmental factor, this was historically taken to suggest that it is also species-typical and functional. Likewise, traits that are species-typical were expected to be developmentally buffered and functional, and so on. However, we now know that these three properties often come apart in nature. Just consider the Bluehead Wrasse (*Thalassoma bifasciatum*), which routinely changes sex from male to female (and back again) in response to the ratio of males to females in its social environment (Warner and Swearer, 1991). Such examples illustrate that a species-typical and (presumably) functional trait can be highly phenotypically plastic (not fixed). Likewise, other traits which are more developmentally stable, such as song-structure in some birds, are variable across a species' range (Benedict and Bowie, 2009). Other behaviors are induced by exposure to certain chemicals during development, and could become typical without being functional (Zala and Penn, 2004). A problem with the vernacular concept of innateness is that it promotes unreliable inferences from one property to another.

Numerous scientists engaged in the study of behavioral development argued that the concept of innateness skews our understanding of development and have called on scientists to abandon it (e.g., Lehrman, 1953; Hinde 1968; West, King, and Duff, 1990), a call endorsed by some philosophers (Griffiths, 2002; Mameli and Bateson, 2006; Stotz, 2008; Griffiths and Machery, 2008). Despite this criticism, the concept of innateness is alive and well in many areas of science. Mainstream developmental psychologists theorize regularly about whether a trait is innate or about the innate foundations of the mind. In her influential book, *The Origin of Concepts*, Susan Carey writes as follows (2009, p. 67; emphasis added):

> I agree with these writers [Renée Baillargeon, Randy Gallistel, Rochel Gelman, Alan Leslie, and Elizabeth Spelke] that the cognition of humans, like that of all animals, begins with highly structured *innate* mechanisms designed to build representations with specific content.

Similarly, linguists of a Chomskian persuasion hold that people can learn to speak a language because of the innate endowment of the mind. Baker describes this psychological endowment as follows (2001, p. 13; emphasis added):

> By their very nature, children seem to be especially equipped for language learning. No one yet knows exactly what this special equipment consists of. It probably involves knowledge of what human languages are like and of what kinds of sounds and structures they might contain, together with strategies for recognizing those sounds and structures. Linguists call this *innate* head start "universal grammar."

Evolutionary behavioral scientists also appeal to the notion of innateness. Boyd and Richerson refer to the "*innate* predispositions and organic constraints that influence the ideas that we find attractive, the skills that we can learn, the emotions that we can experience, and the very way we see the world" (2005, p. 8; emphasis added). And so do many philosophers, such as Fodor in his classic article "The present status of the innateness controversy" (1981, 258; emphasis added):

> It does seem to me that Chomsky's demonstration that there is serious evidence for the *innateness* of what he calls "General Linguistic Theory" is the existence proof for the possibility of a cognitive science.

Importantly, it is not only nativists who use the notion of innateness in a scientific context. Empiricists, who assert that many traits are not innate but acquired, often use this notion too. For instance, Prinz describes his target in *Beyond Human Nature* as follows (2012, p. 2; emphasis added):

> By ignoring cultural variation, researchers end up giving us a misleading picture of the mind. We end up with the idea that psychology is profoundly inflexible. This outlook grossly underestimates human potential. It leads to the view that our behavior is mostly driven by biology. Mainstream cognitive scientists give the impression that human traits are "innate", "genetic" or "hardwired."

Why do these scientists and philosophers adhere to a concept[1] that has been so thoroughly criticized? It may be that while scientists, like laypeople, use the word *innate*, they do so in order to express a different, less objectionable (if at all) concept of innateness. The term *innateness* is preserved because the new, scientific, defensible concept of innateness is a revision of the problematic concept of innateness: It is a "successor concept" (Machery, 2017a). An alternative hypothesis is that some researchers cling to this concept for less epistemically justified reasons. It is possible that even sophisticated researchers occasionally revert, perhaps unwittingly, to using the folk concept. On this view, there is no revised conception of innateness that scientists share. Rather, there is a prescientific folk concept that occasionally influences their thinking, presumably in ways that are misleading.

The goal of this chapter is to investigate what scientists mean by *innateness* in order to find out whether they have developed a successor concept or whether they are still relying on the vernacular concept of innateness. To meet this goal, it is not sufficient to ask scientists what they mean by *innate* or to report their explicit theorizing about innateness. Concerning the former point, scientists may have false beliefs about the very concept of innateness they happen to use; indeed, they are probably incentivized to claim that they mean something different from what laypeople mean. Concerning the latter point, scientists may have developed a particular understanding about what innateness is, but this understanding is not at work in their everyday use of the term *innate*: In their most reflective moments, scientists may mean one thing, while using another concept in their everyday judgments about innateness. We will call the concept of innateness scientists happen to use when they judge that something is or isn't innate their *effective* concept and the concept that is expressed when they theorize explicitly about innateness their *explicit* concept. Using this terminology, our concern then is that scientists' effective concept might not be their explicit concept.

How can we identify the characteristics of scientists' effective concept of innateness? Instead of asking scientists what they mean and instead of looking at their writings about innateness, we propose to examine how they use the word *innate* and infer the content of the expressed concept from these uses (Hyundeuk and Machery, 2016; Machery, 2017b). To do so, we could examine the natural occurrences of this word in scientists' writings, as do linguists working with corpora, but this approach has well-known limitations (Griffiths and Stotz, 2008). Alternatively, we can elicit uses of *innate* and infer scientists' concepts from these uses. In this chapter, we will follow this experimental methodology in order to characterize the content of scientists' effective concept of innateness.[2]

Here is how we proceed. In the first section titled "The vernacular concept of innateness," we review the experimental work on the vernacular concept of innateness, which will form the background of the experimental work reported in this chapter. In "Two hypotheses about scientists' concept of innateness," we present two distinct hypotheses about scientists' concept of innateness: the conceptual ecology hypothesis and the attractor hypothesis. The section "Experimental study" describes our experimental work and our results. The next section "Scientific and folk concepts" discusses the significance of these results for the relation between folk and scientific concepts. Finally, "Innateness in science" delves into the significance of our project for the concept of innateness in science.

The vernacular concept of innateness

Following Griffiths (2002), we proposed in past research that the vernacular concept of innateness belongs to folkbiology, and we used psychological research on folkbiology to formulate hypotheses about the content of the concept of innateness (Griffiths, Machery, and Linquist, 2009; Linquist et al., 2011). An important component of folkbiology is the belief that biological traits are inherited at birth from genetic parents and that their development is relatively insensitive to the offspring's environment. The switched-at-birth experimental task has been used to examine this belief in children (e.g., Gelman, 2004): In this task, a vignette describes how a young animal (e.g., a young cow) is raised by members of another species (e.g., by pigs), and children are asked whether the animal would have the traits of its genetic parents (e.g., they are asked whether the young cow would moo) or of its adoptive parents. (This task has also been adapted to describe a child being reared by parents different from its genetic parents in various respects.) Research suggests that biological traits are thought to be transferred at birth, independently of the rearing environment. On this basis, we hypothesized that when laypeople say that a trait is innate, they view it as developing reliably independently of its environment. Psychologists have also emphasized the role of generics in folkbiology (e.g., Gelman, Ware, and Kleinberg, 2010). Generics are statements that, while expressing generalizations, do not contain quantifiers, such as "all" or "every." While tolerating exceptions, they are often understood as asserting that a property is widely shared for nonaccidental reasons by the members of a given class. Lay claims about species are often expressed by means of generics. On the basis of this body of research, we hypothesized that when laypeople say that a trait is innate, they view it as widespread among conspecifics. Finally, psychologists have identified a form of naïve teleology at work in lay thought (Kelemen and Rosset, 2009): laypeople ascribe functions to biological traits (as well as to other natural things). On the basis of this body of evidence, we hypothesized that when laypeople say that a trait is innate, they view it as having some kind of function. To summarize, we formulated three hypotheses about the content of the concept of innateness:

1. Fixity: A trait is more likely to be judged innate the more independent from its environment its development appears to be.
2. Typicality: A trait is more likely to be judged innate the more typical it is.
3. Teleology: A trait is more likely to be judged innate the more functional it is.

To clarify, we do not view fixity, typicality, and teleology as separately necessary and jointly sufficient conditions for a trait to count as innate. Rather, we view them as prototypical features of the concept of innateness; they characterize what a prototypical innate trait would look like, and a trait is more likely to be judged innate to the extent that it resembles this prototypical innate trait.[3]

To test our hypothesis about the vernacular concept of innateness, we developed a set of eight vignettes that varied whether a trait (birdsong) develops reliably, is typical, and has a function. Participants were then asked whether the trait is innate. Here is one of the vignettes we used:

> Birdsong is one of the most intensively studied aspects of animal behaviour. Since the 1950s scientists have used recordings and sound spectrograms to uncover the structure and function of birdsong. Neuroscientists have investigated in great detail the areas of the brain that allow birds to develop and produce their songs. Other scientists have done ecological fieldwork to study what role song plays in the lives of different birds.
>
> The Alder Flycatcher (*Empidonax alnorum*) is a migratory neo-tropical bird which breeds in southern Canada and the northern USA. Studies on the Alder Flycatcher show that the song an adult male produces does not depend on which songs they hear when they are young. Studies also show that different males in this species sing different songs. Furthermore, close observations of these birds reveal that the males' song attracts mates and helps to defend their territory. Scientists therefore agree that the bird's song has a real function, like the heart in humans.

On a 7-point scale, 1 meaning strongly disagree and 7 meaning strongly agree, how would you respond to the following statement?

> "The song of the male Alder Flycatcher is innate."

The seven other vignettes were variations based on this vignette. Participants were laypeople without any training in biology or psychology. In Study 1, they were presented with a single vignette (between-subjects design), while in Study 2, they were presented with the eight vignettes (within-subjects design).

While there were a few differences between the results of these two studies, these were largely congruent: Fixity and typicality had a strong influence on how strongly participants agreed that a trait is innate (how much they matter depends on the details of the experimental design), while, surprisingly, teleology had no significant impact on participants' judgments, at least not one that was detectable using our questionnaire. Importantly, the two properties of fixity and typicality independently influenced participants' judgments. For example, if a trait was presented as either fixed (and not typical), or typical (and not fixed), then respondents would be inclined to agree that the trait is innate in either case. We took this to reveal something important about the structure of the vernacular concept: that neither fixity nor typicality are regarded as necessary conditions for innateness.

Follow-up work has confirmed and extended these results (Linquist et al., 2011; see also Knobe and Samuels, 2013). One of the main goals of this follow-up work was to examine whether other expressions in contemporary English express the vernacular concept of innateness (or very related ones). Lay participants were presented with a set of eight vignettes based on the vignettes previously used. However, instead of using lay terminology, the vignettes involved more accurate, technical terminology. For instance, once of the vignettes read as follows:

> Sarkar's Sparrow (*Aimophila sarkarii*) is one of the many species of American sparrow. It is found in Mexico and southwest Texas. Historically, it was more widely distributed in the southwestern USA, but its range has contracted as a result of overgrazing by livestock. It can be shown by experimentally manipulating what young birds hear that *the sequence of song elements* produced by an adult Sarkar's Sparrow male depends on which sequences it hears when it is young. Furthermore, studies have shown that there is significant interpopulational and interindividual variation in the sequence of song elements produced by Sarkar's Sparrow males. Finally, close observations of these birds reveal that the sequence of song elements produced by Sarkar's Sparrow males does not help them to attract mates and does not help them to defend their territory. Scientists therefore agree that the sequence of song elements produced by Sarkar's Sparrow males is not an adaptation.

Each participant saw four vignettes (four vignettes where the song is fixed or four vignettes where it is not fixed). Participants were asked the following question followed by one of the three following statements:

> On a 7-point scale, 1 meaning strongly disagree and 7 meaning strongly agree, how would you respond to the following statement?
> The sequence of song elements produced by a male [species name] is innate.

The sequence of song elements produced by a male [species name] is part of its nature.

The sequence of song elements produced by a male [species name] is in its DNA.

Answers to the innateness question replicated by and large the results of Griffiths, Machery, and Linquist, 2009. More interesting, answers to the "in its DNA" question matched our expectations about the concept of innateness extremely well, while answers to the "in its nature" question were hard to interpret. We interpret these results to suggest that laypeople use "innate" and "in the genes" to express the same, vernacular concept. More generally, the results help to establish the robustness of our earlier finding that the vernacular concept of innateness is well characterized by the three hypotheses we derived from the research on folkbiology (although teleology is less important than expected). In everyday English, the often used expression "it is in their genes" expresses this concept and draws a distinction between the traits that reflect the nature of an animal or a plant and the traits that don't. The expression "in their nature" is not very clear for contemporary speakers of English, while "innate" roughly maps onto the concept of innateness.

Our three-factor model is also convergent with a large body of work on genetic essentialism conducted in social psychology at around the same time (e.g., Haslam and Ernst, 2002; Dar-Nimrod and Heine, 2011; Haslam, 2011; Cheung, Dar-Nimrod, and Gonsalkorale, 2014). The congruence between the results of these two bodies of work can be seen in Table 8.1 (see also Stotz and Griffiths, 2018; Lynch et al., 2019).

Two hypotheses about scientists' concept of innateness

Earlier, we sketched two hypotheses to explain scientists persistent use of *innate* and other cognate terms, despite the widespread recognition that the vernacular concept is scientifically flawed. The first proposal stated that the vernacular concept has effectively evolved and become adapted to the particular demands of contemporary science. The alternative proposal stated that researchers are effectively stuck in an outmoded way of thinking, which is potentially counterproductive. In this section, we refine these hypotheses and draw connections to the literature on scientific theory change and cultural evolution. Both hypotheses agree that scientists' concept or concepts of innateness are

Table 8.1 Comparison between the genetic essentialism framework (GEF) and the three-factor model (from Stotz and Griffiths 2018, p. 61)

Genetic Essentialist Elements	Three-Factor Model of Animal Natures
(Dar-Nimrod and Heine, 2011)	(Linquist et al., 2011)
Immutable and determined: thinking about genetic attributions leads people to view relevant phenotypes as less changeable and predetermined	*Fixity*: phenotypes that are part of an animal's nature do not depend on the particular environment in which the organism is raised and are hard to change by environmental manipulations
Specific etiology: the tendency to discount additional causal explanations once genetic attributions are made	Traits are *either* expression of the animal's nature (and are expected to have the three features) *or* imposed by the environment (with opposite expectations)
Homogeneous and discrete: leads to a focus on the central identifying features that are common to all group members, drawing attention away from in-group differentiating features	*Typicality*: phenotypes that are part of an animal's nature are typical of the entire species or of some natural subset such as males or juveniles
Nature: phenotypes are perceived as a natural outcome (with positive normative associations)	*Teleology*: phenotypes that are part of an animal's nature serve some purpose (with positive normative associations)

derived from the vernacular concept of innateness. Both hypotheses also assume that the research summarized in "The vernacular concept of innateness" captures the important features of the vernacular concept of innateness. They disagree, however, about the relation between the vernacular concept of innateness and scientists' concept or concepts of innateness.

The first approach is inspired by Hull's (1988, Chapter 12) evolutionary philosophy of science, and an approach which Stotz and Griffiths (2004) have termed "conceptual ecology." According to Hull, theories evolve and are connected to one another by a process that shares many similarities with evolution by natural selection. Theories are related to one another by a process of descent, theories typically being modifications of preexisting theories. Not all modifications are successful: Some modifications allow a theory to better fit with its niche, where the niche of a theory includes the set of explananda it is brought to bear on, the interventions it allows scientists to make, and so on. Stotz and Griffiths (2004, see also Griffiths and Stotz, 2008) have conducted research in the same spirit, examining the "ecological" context of scientific concepts ("conceptual niches"). On their view, scientists modify their conceptual tools flexibly and adaptively to respond to theoretical and experimental needs, which vary across research communities and traditions. There are differences between these two approaches to scientific dynamics (for one, Hull focuses on theories, Stotz and Griffiths, on concepts), but both have the same consequence for scientists' concepts of innateness: We should expect variation across disciplines in how scientists conceptualize innateness. All these concepts derive from the vernacular concept of innateness (the ancestral concept) and have come to fit their particular scientific niche. We can then formulate the conceptual ecology hypothesis:

> *The conceptual ecology hypothesis*: The concept of innateness should vary across scientific disciplines or scientific traditions, depending on varying theoretical, explanatory, experimental, and practical needs.

The alternative attractor hypothesis is loosely inspired by Sperber's theory of cultural evolution: cultural attractor theory (Sperber, 1996; Claidière and Sperber, 2007). According to Sperber, what beliefs, concepts, values, desires, and so on people happen to embrace in a given culture is influenced and constrained, among other things, by evolved, universal psychological structures. This influence explains the similarities across cultures. We propose to extend some elements of this theory of cultural evolution to scientific concepts. We propose that the vernacular concept of innateness works as an attractor: Scientists tend to express this very concept by the word *innate*. So to speak, their thoughts are attracted by this concept, and it is difficult for them to develop an effective concept that differs from it. We can now formulate the attractor hypothesis:

> *The attractor hypothesis*: Scientists tend to think similarly about innateness across disciplines or scientific traditions because the vernacular concept of innateness influences their thoughts on the matter.

Thus, we must consider two competing hypotheses about the relation between the vernacular concept of innateness and scientists' concept(s) of innateness. In the remainder of this chapter, we test these two hypotheses experimentally.

Experimental study

Participants

Participants were scientists working in linguistics, psychology (including comparative, developmental, and evolutionary psychology), neuroscience, genetics, biology (including developmental and evolutionary biology), and ethology. In September 2010, in collaboration with Joshua Knobe and Richard

Samuels the coauthors of this paper contacted hundreds of scientists by email. The email we sent asked scientists to log in into a Qualtrics website in order to complete a short survey; they were also asked to forward the email to their colleagues and graduate students. In response to this email, 767 participants completed the survey. Out of those, 295 were randomly assigned to the study reported in this chapter. The remainder completed the studies conducted by Knobe and Samuels (reported in Knobe and Samuels, 2013). We excluded participants who did not complete the survey, leaving us with a sample of 232 scientists (Table 8.2).

Materials and procedures

Eight vignettes were constructed on the basis of our previous work on the vernacular concept of innateness (Griffiths et al., 2009; Linquist et al., 2011). Each vignette was about birdsong. The eight vignettes then varied whether birdsong was fixed, typical, and had a function. As in our previous work, we made sure that the content of each vignette was true and seemed plausible since scientists might be reluctant to make a judgment about implausible scenarios. We relied on the more technical sounding and more accurate formulation previously used by Linquist et al. (2011) since participants were scientists and likely familiar with such terminology. As an example, the Eastern Phoebe vignette describes a trait that is fixed, typical, and has a function:

> The Eastern Phoebe (*Sayornis phoebe*) is a small, North American flycatcher. It can be shown by experimentally manipulating what young birds hear that the sequence of song elements produced by an adult Eastern Phoebe male does not depend on which sequences it hears when it is young. Furthermore, studies have shown that there is no significant interpopulational and interindividual variation in the sequence of song elements produced by Eastern Phoebe males. Finally, close observations of these birds reveal that the sequence of song elements produced by Eastern Phoebe males helps them attract mates and helps

Table 8.2 Participants' demographics

Gender	Language	Nationality (main countries)	Education
Male: 48%	English: 72%	USA: 55.7% Canada: 7.3% UK: 5.5% Germany: 4.8%	College degree: 17:3% MA: 24.3% PhD: 58.4%

them to defend their territory. Scientists therefore agree that the sequence of song elements produced by Eastern Phoebe males is an adaptation.

Participants were first presented with an accurate introduction about scientific research on birdsong:

> Birdsong is one of the most intensively studied aspects of animal behaviour. Since the 1950s scientists have used recordings and acoustic analysis to uncover the structure and function of birdsong. Neuroscientists have investigated in great detail the areas of the brain that allow birds to develop and produce their songs and hormonal influences on song production. Other scientists have done ecological fieldwork to study what role song plays in the lives of different birds.

They were then presented with four out of the eight possible vignettes (all of them describing either a fixed birdsong or a nonfixed birdsong). The order of these four vignettes was randomized. Each vignette was followed by the following question:

> On a 7-point scale, 1 meaning strongly disagree and 7 meaning strongly agree, how would you respond to the following statement?
> "The sequence of song elements produced by a male Black-Capped Chickadee is innate."

Participants were asked to respond on a 7-point scale, anchored at 1 with "Strongly disagree" and at 7 with "strongly agree." This mixed factorial design (between-subjects factor: fixity; within-subjects factor: typicality and teleology) follows Linquist et al., 2011.

Participants then completed a standard demographic questionnaire as well as a questionnaire reporting their disciplinary affiliation. Participants were allowed to check several options (e.g., developmental psychology and evolutionary psychology). We use their answers to this latter questionnaire in the analysis reported below. Finally, participants were asked to answer the question, "In your view, does the notion of *innateness* have any role within your field(s) of research?" on a 7-point scale anchored at 1 with "No," 4 with "In between," and 7 with "Yes."

Results

Our first task was to assign participants to disciplines. Because participants were allowed to check several options, we had to assign some of them to one of the options they had chosen. Before analyzing the data, we adopted the following rules:

1. Participants are classified as psychologists if they checked one of the psychology options (except for the "other" option, which was treated on a case-by-case basis).
2. Participants are classified as human behavioral scientists if they checked "evolutionary psychology" or "human behavioral ecology."
3. Participants are classified as experimental psychologists if they checked "social," "experimental," or "personality" psychologists and are not classified as evolutionary behavioral scientists.
4. Participants are classified as linguists if they checked one of the linguistics options (except for the "other" option, which was treated on a case-by-case basis) and if they were not classified as psychologists.
5. Participants are classified as generative linguists if they checked the "generative linguistics" option.
6. Participants are classified as nongenerative linguists if they checked the "sociolinguistics" option.
7. Participants are classified as biologists if they checked one of the biology options (except for the "other" option, which was treated on a case-by-case basis).

In addition, we used participants' answers to the "other" option to classify a few additional participants in these categories.[4] Psycholinguists, anthropologists, and ethologists were classified as psychologists. Cultural anthropologists were classified as experimental psychologists. Evolutionary anthropologists were classified as evolutionary behavioral scientists. Cognitive linguists as well as linguists working on typology, applied linguistics, comparative linguistics, and documentary linguistics were added to the nongenerative linguistics category.

We relied on this disciplinary classificatory scheme for several reasons. First, we intended to compare scientists' concepts of innateness across broad disciplinary distinctions: The contrast between psychology, linguistics, and biology is meant to allow for this comparison. Second, within each of these broad disciplines, researchers within subdisciplines or competing approaches either have expressed very different views about the innate endowment of the mind or rely on the concept of innateness to a very different extent. Evolutionary psychologists use the concept of innateness extensively, while cognitive psychologists, social psychologists, and personality psychologists much less. Generative linguists tend to believe that key components of language are innate, while sociolinguists are skeptical. Finally, we were constrained by our small sample size: Many possibly interesting disciplinary groupings resulted in sample sizes that were too small to allow for meaningful quantitative analysis.

We first examine how scientists conceive of innateness across the three broadest disciplinary affiliations we distinguished. A mixed-design analysis of psychologists' responses (between-subjects factor: fixity; within-subjects factors: typicality and teleology) revealed a main effect of typicality (F(1, 78) = 24.2, $p < .001$; $\eta^2 = .24$) and fixity (F(1, 78) = 47.6, $p < .001$; $\eta^2 = .38$). The effect of typicality was significantly larger when the trait was fixed than when it was not fixed, although the effect size is small (F(1, 78) = 4.6, $p < .03$; $\eta^2 = .06$). Teleology was not significant (F(1, 78) = 2.9, $p = .09$; $\eta^2 = .04$). Figure 8.1 represents these results.

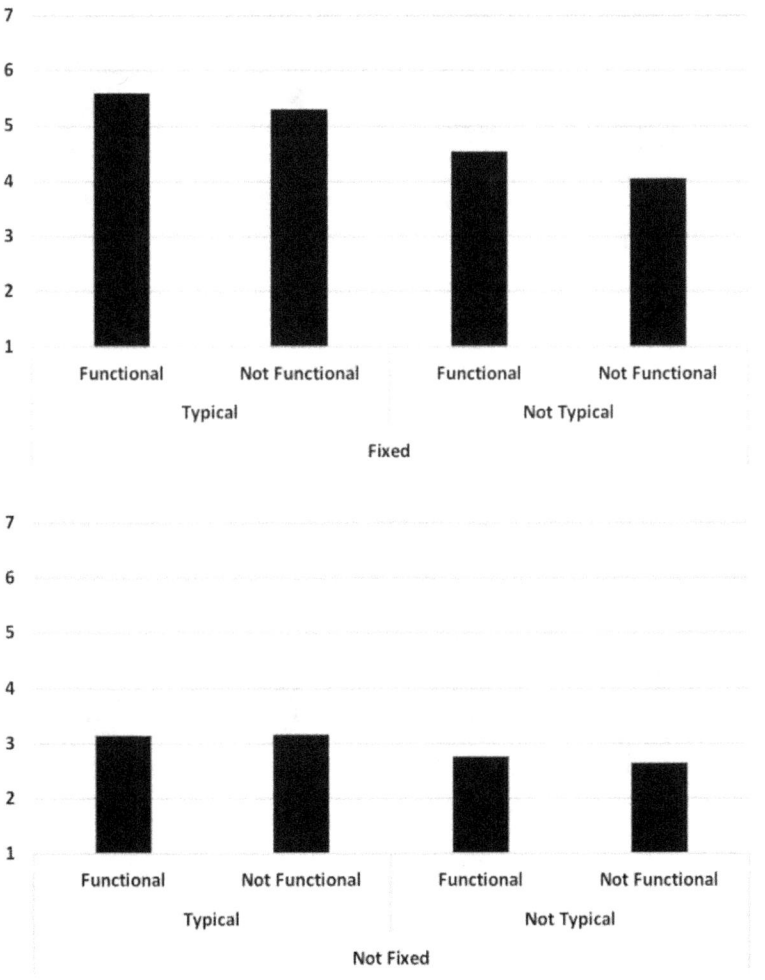

Figure 8.1 Psychologists' mean innateness judgments.

A mixed-design analysis of linguists' responses produced similar results. It revealed a main effect of typicality (F(1, 46) = 11.2, p = .002; η^2 = .20) and fixity (F(1, 46) = 46.9, p < .001; η^2 = .51). No interaction between typicality and fixity was observed. Teleology was not significant (F(1, 46) = 1.4, p = .24; η^2 = .03). Figure 8.2 represents these results.

A mixed-design analysis of biologists' responses produced again similar results. It revealed a main effect of typicality (F(1, 24) = 8.0, p = .009; η^2 = .25) and fixity (F(1, 24) = 11.8, p = .002; η^2 = .33). Teleology was not significant (F(1, 24) = .5, p = .51; η^2 = .02). Figure 8.3 represents these results.

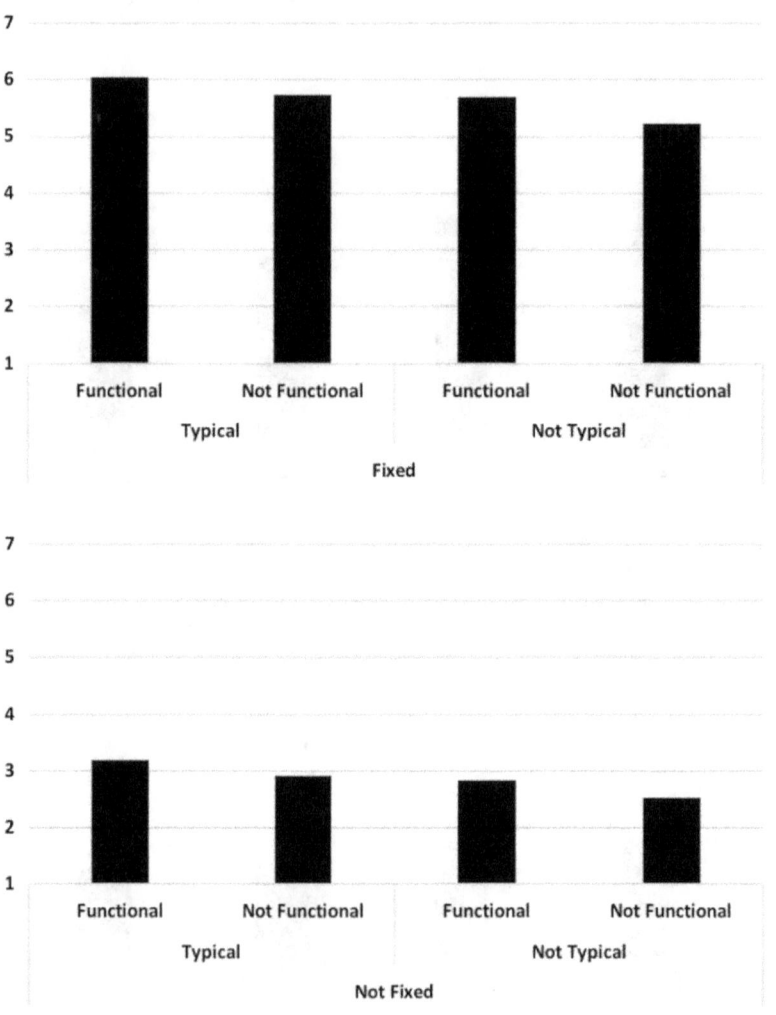

Figure 8.2 Linguists' mean innateness judgments.

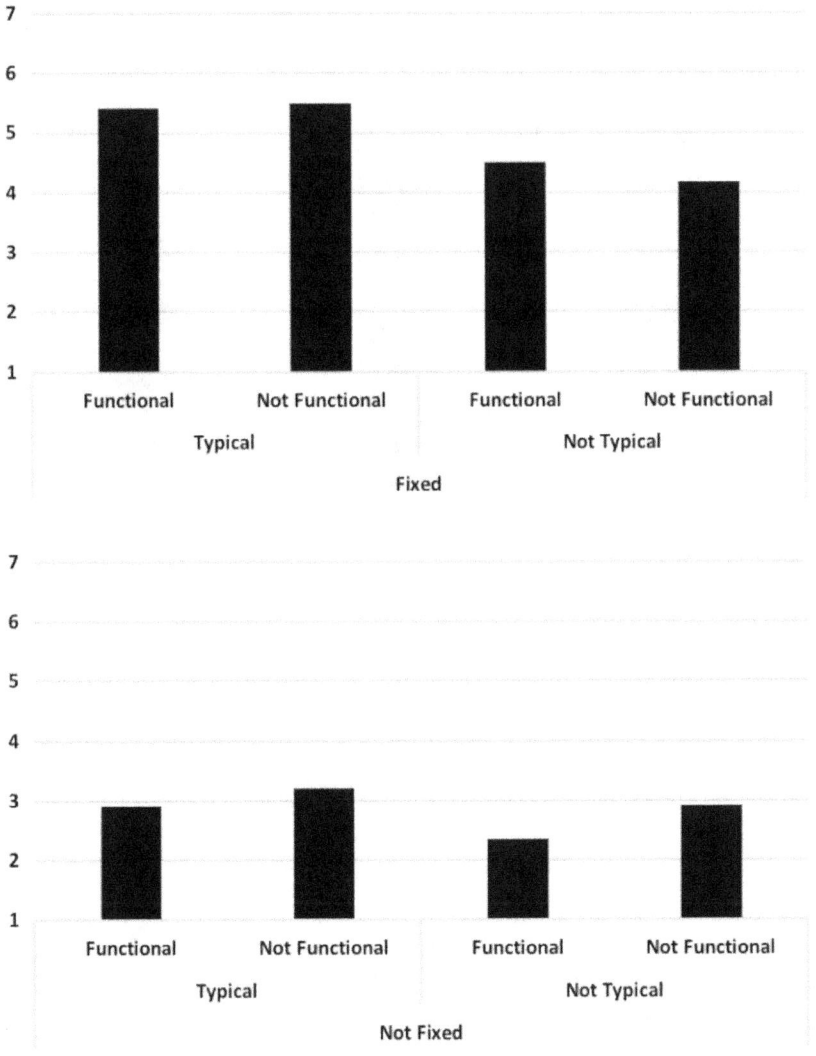

Figure 8.3 Biologists' mean innateness judgments.

We then compared whether the concept of innateness plays a role in research across these three areas. A one-way analysis of variance (ANOVA) established that the concept of innateness is not equally important in biology, linguistics, and psychology (F(2, 24) = 6.3, p = .002; η^2 = .08), with biology being significantly lower than both linguistics and psychology (Figure 8.4).

T-tests revealed that the mean response is significantly higher than the midpoint (anchored at "in between" between "no" and "yes") for both

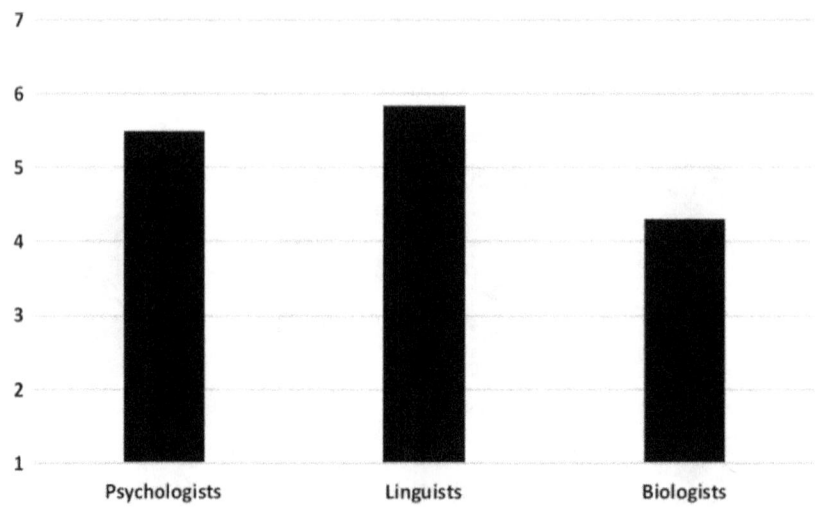

Figure 8.4 Judgments about the importance of innateness across disciplines.

psychologists ($t(76) = 7.2, p < .001$) and linguists ($t(46) = 8.0, p < .001$), but not for biologists ($t(26) = .7, p = .5$).

We now turn to some of the subdisciplines identified earlier. A mixed-design analysis of evolutionary behavioral scientists' responses produced again the same pattern of results. It revealed a main effect of typicality ($F(1, 30) = 4.4, p = .045; \eta^2 = .13$) and fixity ($F(1, 30) = 26.0, p < .001; \eta^2 = .46$). Teleology was not significant ($F(1, 30) = 2.5, p = .12; \eta^2 = .08$). Figure 8.5 represents these results.

A mixed-design analysis of experimental psychologists' responses produced a similar pattern of results. It revealed a main effect of typicality ($F(1, 33) = 19.3, p < .001; \eta^2 = .37$) and fixity ($F(1, 33) = 21.6, p < .001; \eta^2 = .40$). The main effect of typicality was qualified by an interaction with fixity ($F(1, 33) = 10.6, p = .003; \eta^2 = .24$). Teleology was not significant ($F(1, 33) = .13, p = .72; \eta^2 < .01$). Figure 8.6 represents these results.

Evolutionary behavioral scientists gave marginally more positive answers than experimental psychologists to the question, "Does the notion of innateness have any role within your field(s) of research?" ($t(64) = 1.98, p = .052$; Figure 8.7) but both groups were more likely to give a positive answer compared to midpoint (evolutionary behavioral scientists: $t(30) = 7.3, p < .001$; experimental psychologists: $t(34) = 2.9, p = .007$).

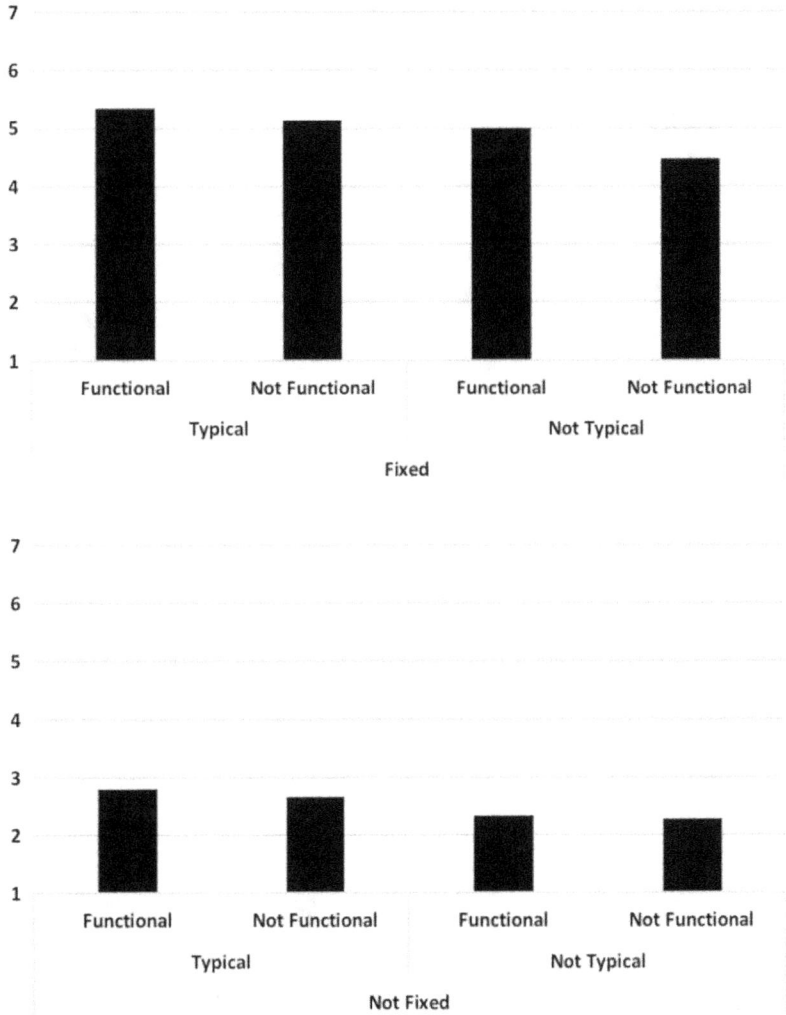

Figure 8.5 Evolutionary behavioral scientists' mean innateness judgments.

A mixed-design analysis of generative linguists' responses produced the usual pattern of results. It revealed a main effect of typicality (F(1, 20) = 12.4, p = .002; η^2 = .38) and fixity (F(1, 10) = 8.8, p = .008; η^2 = .31). Teleology was not significant (F(1, 20) = 1.4, p = .25; η^2 = .07). Figure 8.8 represents these results.

A mixed-design analysis of nongenerative linguists' responses produced a similar pattern of results. It revealed a main effect of typicality (F(1, 24) = 9.8,

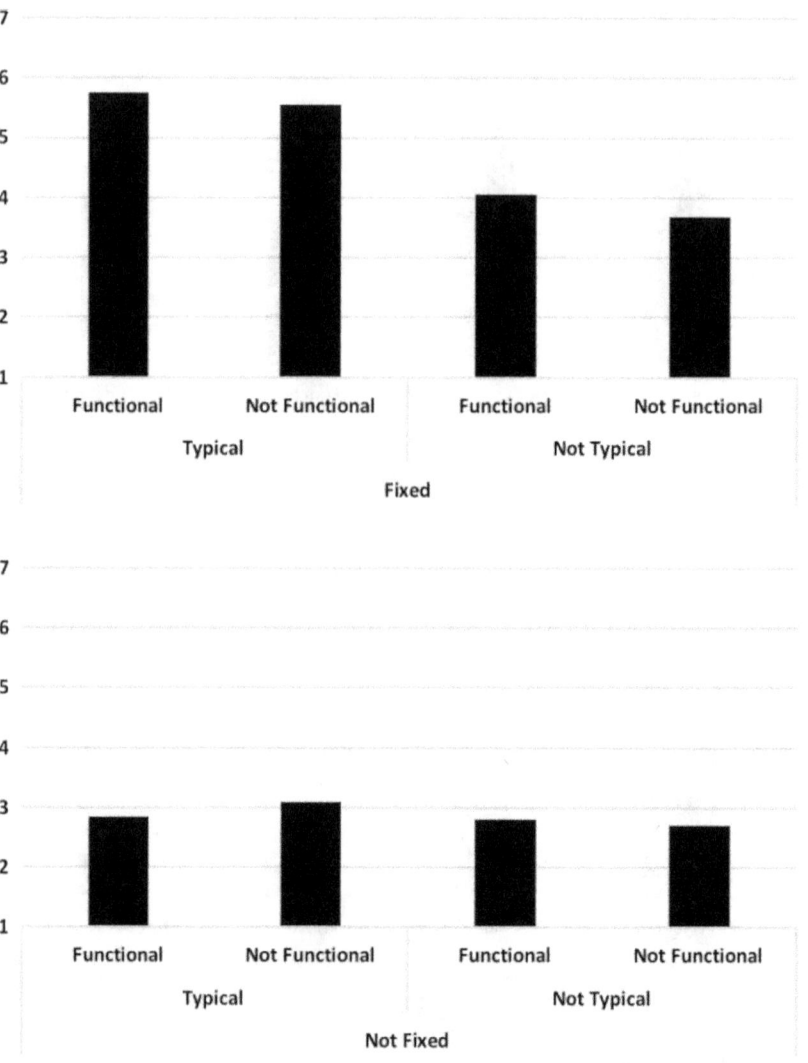

Figure 8.6 Experimental psychologists' mean innateness judgments.

$p = .004$; $\eta^2 = .29$) and fixity (F(1, 24) = 32.6, $p < .001$; $\eta^2 = .58$). Teleology was not significant (F(1, 24) = .74, $p = .40$; $\eta^2 = .03$). Figure 8.9 represents these results.

Generative linguists gave more positive answers than nongenerative linguists to the question, "Does the notion of innateness have any role within your field(s) of research?" (t(46) = 3.7, $p = .001$; Figure 8.10) but both groups were more likely to give a positive answer compared to midpoint (generative linguists: t(21) = 16.1, $p < .001$; experimental psychologists: t(25) = 3.4, $p = .002$).

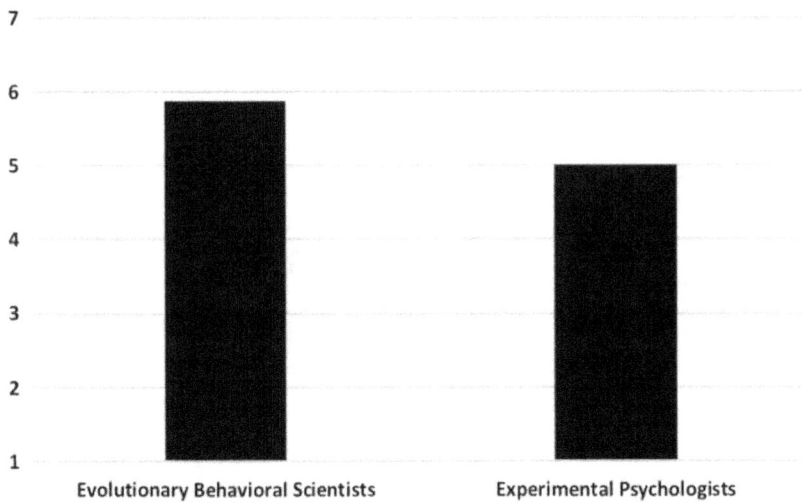

Figure 8.7 Judgments about the importance of innateness among psychologists.

Limitations

Before discussing the significance of the results reported here, it is worth emphasizing their limitations. First, the data were obtained by snowball sampling, and there is no guarantee that our sample is representative of the broader community of linguists, psychologists, and biologists. Second, the sample size is small, particularly when it comes to the subdisciplines we examined. The study is thus low powered, and negative results are difficult to interpret. Third, the classificatory scheme we used was imperfect; participants were allowed to choose several options, and we had to interpret their answers. The rules were developed in advance of analyzing the data, but they were not preregistered and left room for judgment calls. Fourth, our sample is mostly American, raising questions about its generalizability to other countries. For these four reasons mainly, this study of scientists' concepts of innateness remains exploratory. That said, even an exploratory study with such limitations is better than evidence-free speculations about scientists' concepts of innateness.

Discussion

Earlier, we compared two hypotheses about the transformation of a vernacular concept into a scientific concept: the conceptual ecology and the attractor hypotheses. Each hypothesis postulates its own model of the psychological

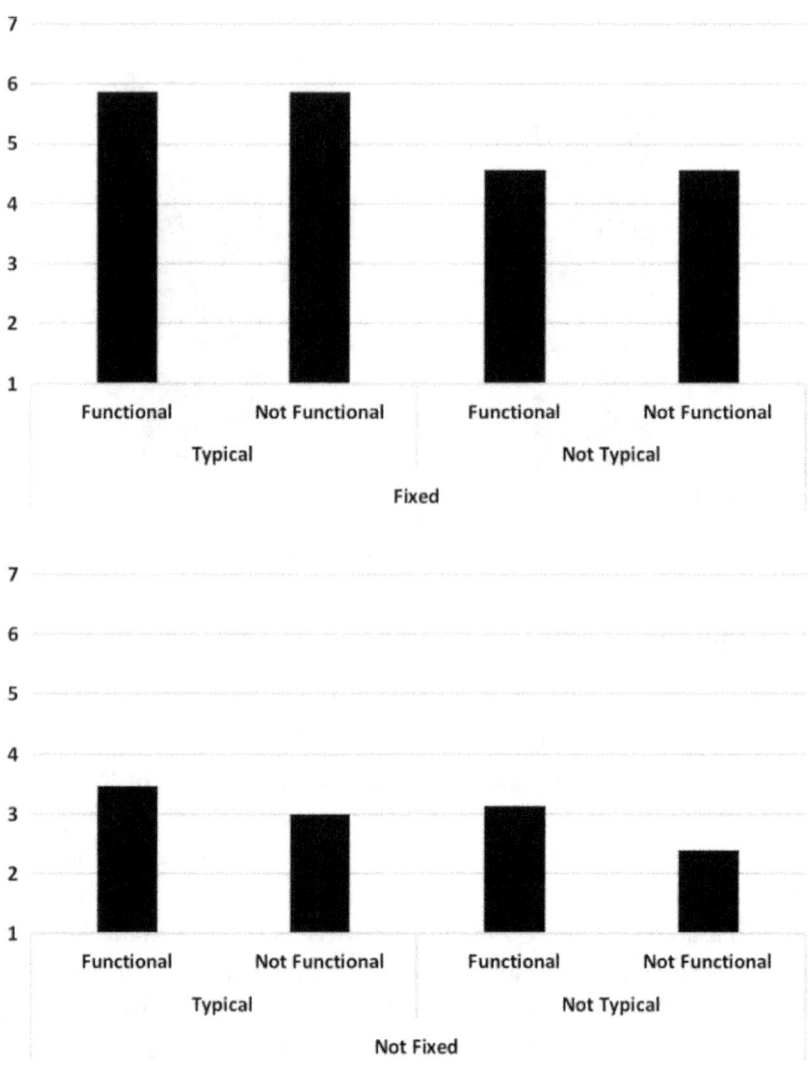

Figure 8.8 Generative linguists' mean innateness judgments.

factors influencing the changes in a concept when it becomes part of scientists' toolkit; it also postulates different trajectories in its transformation into a scientific concept. Most relevant here, they make different predictions about the similarities and differences across scientific communities: The former hypothesis predicts diversity among scientific communities as well as difference with the vernacular concept of innateness, the latter similarity among scientific communities as well as with the vernacular concept of innateness.

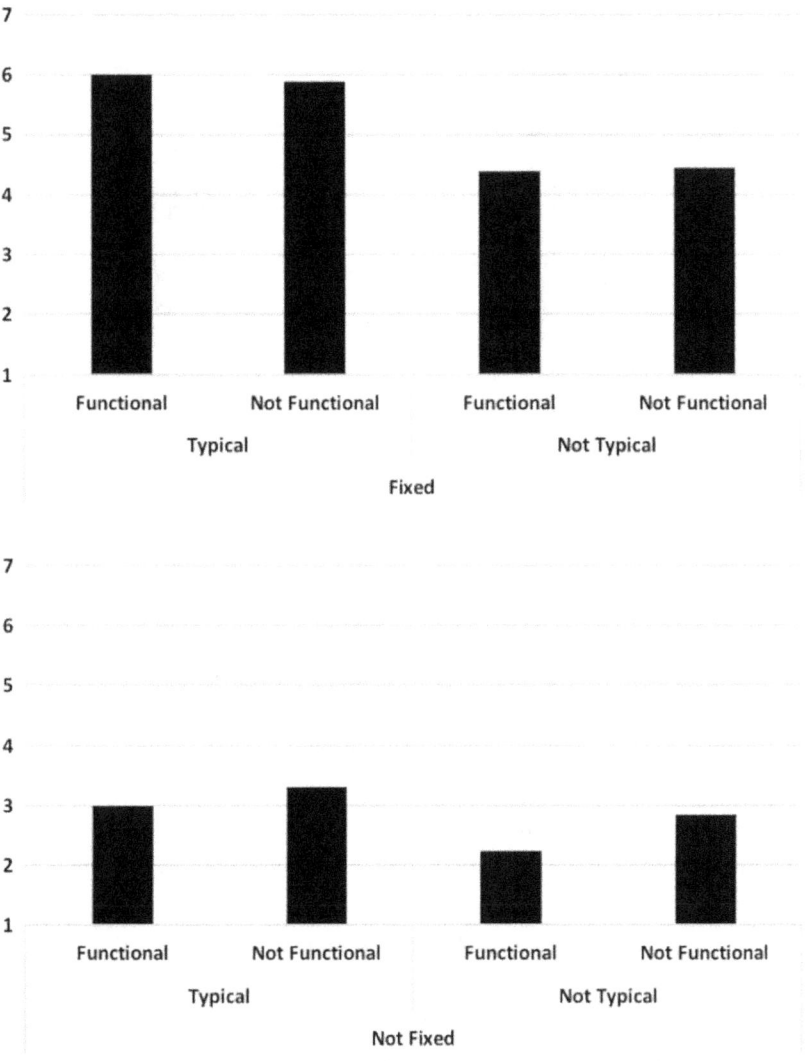

Figure 8.9 Nongenerative linguists' mean innateness judgments.

Our data fit the attractor model much better than the conceptual ecology model. While there may be some small differences across scientific disciplines and between scientists' concept of innateness and the vernacular concept of innateness (more on this below), the similarities are striking. In all the disciplines and subdisciplines we examined, typicality and fixity were important contributors to judgments about innateness, as is the case for laypeople (Griffiths, Machery, and Linquist, 2009; Linquist et al., 2011). Teleology was

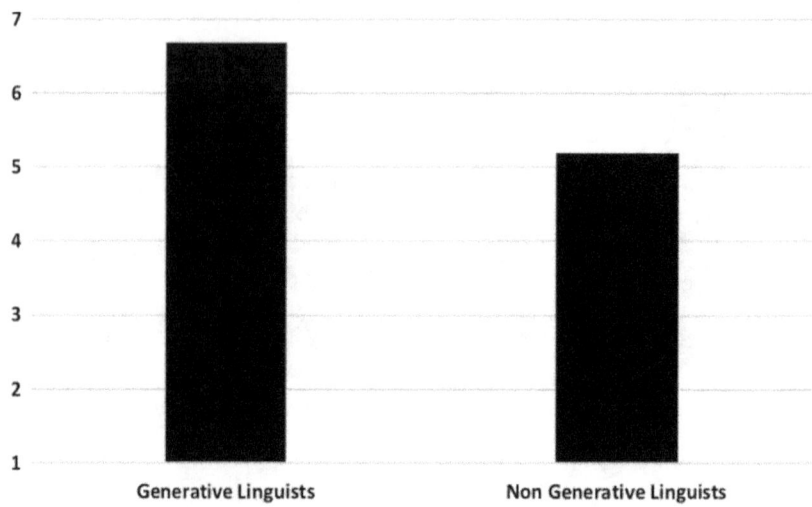

Figure 8.10 Judgments about the importance of innateness among linguists.

not a significant predictor of judgments about innateness; if it predicts them, its influence is small and cannot be detected with the power of our study. This too is in line with our previous results: Griffiths, Machery, and Linquist, (2009) found a small effect of teleology on judgments about innateness, but this effect was not detected by Linquist et al. (2011).

Three results should be highlighted. We didn't find any difference between generative and nongenerative linguists: These two communities sharply disagree about the extent to which language is innate, but they conceive of innateness in a very similar way. Human behavioral scientists did not seem particularly sensitive to the function of a trait when deciding whether it is innate, despite often endorsing an adaptationist perspective on the study of human behavior. Despite the diversity in their training, theoretical knowledge, and explanatory needs, biologists, linguists, and psychologists apparently did not differ in how they conceive of innateness, at least with respect to the probes used in this study.

We did find two small, hard-to-interpret differences. While typicality and fixity are additive factors for laypeople and for most scientists, we found an interaction effect among psychologists who do not embrace an evolutionary perspective on human behavior. Because the effect was small, we refrain from interpreting it. The second difference is between the relative weight of typicality and fixity across disciplines and in comparison to the vernacular concept, as measured by the respective effect sizes. The mixed-factor design is similar to the one followed in Linquist et al. (2011), where we found that fixity and typicality had roughly the same weight for laypeople. Here we find, on the contrary, that fixity is substantially more influential than typicality for evolutionary behavioral scientists and nongenerative linguists. While the greater weight of fixity may be interpretable in light of the conceptual ecology hypothesis, the explanation is not obvious and would be clearly post hoc.

To our surprise, scientists in all the disciplines judged that the concept of innateness was somewhat important in their discipline. One might have expected at least some groups of scientists to dismiss its role in science, but this is not what we found. We found some variation in scientists' opinions about the importance of this concept: biologists judged it less important than psychologists and linguists; among linguists, generative linguists judged it more important than nongenerative linguists; among psychologists, evolutionary behavioral scientists judged it more important than experimental psychologists.

One may think that our results are partly driven by artificial groupings of scientists. For instance, the category of evolutionary behavioral scientists included both evolutionary psychologists and human behavioral ecologists. These two subcommunities may use the concept of innateness differently. The same can be said to different degrees of our other disciplinary groupings. In response, we concede that this is a limitation of our project. Had our sample been larger, we would have been able to study more natural groupings. On the other hand, the striking uniformity of answers suggests that this objection is ultimately misguided. Disciplinary affiliation just does not seem to matter.

Scientific and folk concepts

Some concepts have a double life: they help laypeople make sense of their everyday interactions with the world, and they are also used in scientific explanations. The concepts of belief, value, desire, norms, motivations, society, language, weight, heat, temperature, and so on illustrate this situation. What happens to such vernacular concepts whey they are embedded in scientific theories and put to scientific uses? One might think that these concepts get transformed by their local scientific uses, either because scientists intentionally transform them (by, e.g., explicitly proposing new definitions) or merely as a result of repeated uses. This process can be observed in the evolution of the concept of the gene over the past century (Griffiths and Stotz, 2013). The results reported here, however, suggest that this outcome is by no means necessary. Rather than being transformed, particularly in functional ways (so as to fit the conceptual and experimental ecologies of the relevant sciences), the concept of innateness is by and large the same across scientific disciplines, and it is very similar to the vernacular concept of innateness. The vernacular concept of innateness leads scientific thinking to slip into a familiar groove (Machery, 2017b, Chapter 7); it is an attractor that prevents scientists from developing locally useful concepts of innateness.

We do not claim that the attractor model always describes the relation between vernacular concepts and scientific concepts; some concepts may be transformed by scientists in a functional manner. But we do claim that it is sometimes an accurate model. So, what explains why in some situations a vernacular concept functions as an attractor while it doesn't in other situations? As this point, we can merely speculate about two factors. First, concepts like that of the gene are tightly anchored to experimental practices and formal schemes of inference (Griffiths and Stotz, 2013). Their formal character severs their use from folk uses. The situation is obviously entirely different for concepts like the concepts of belief, norm, language, and innateness; no formal structure constrains their use, and vernacular concepts can exert their pull. Second, as we noted earlier, the vernacular concept of innateness has long served as an adequate heuristic, informing various sorts of interaction with the biological world. As we saw, it is part of what psychologists call a lay theory, namely, "folkbiology" (Medin and Atran, 1999). It is not a concept acquired by formal learning, which could be displaced by further formal learning. Such concepts are not replaced by scientific concepts; at best, they coexist with them (Knobe and Samuels, 2013; Shtulman, 2017), and they may influence scientists' thinking. Third, while scientists use the concept of innateness often, it may not be explanatorily central to the disciplines under examination. Most of the experiments scientists run and the specific hypotheses they test may not be much affected were this concept removed from scientists' conceptual toolkit. Scientists may be particularly likely to use such concepts loosely and to let vernacular concepts guide their thinking.

Innateness in science

As was discussed earlier, philosophers of biology and psychology have made various claims about how scientists conceive of innateness (for review, see Gross and Rey, 2012; Mameli and Bateson, 2006). Such claims may be true of scientists' explicit characterizations of innateness; they may capture what we called earlier scientists' *explicit* concepts. But our results suggest that they fail to capture their *effective* concept, the concept they deploy when they are asked to think about innateness. Scientists' effective concept is just the vernacular concept of innateness, and, just like laypeople, scientists are more likely to judge a trait to be innate if its development does not appear to depend on aspects of the environment and if it is shared by many conspecifics.

At this juncture, philosophers of biology and psychology who have proposed analyses of scientists' concepts of innateness may respond as follows. They may concede that in our experiments. scientists deploy the vernacular concept of innateness, but they may insist that this isn't scientists' effective concept. Scientists may not have been particularly careful and attentive. After all, the experimental study was not part of their scientific research and was quite different from the kind of context in which they put their scientific concepts to use. So, they defaulted onto the vernacular concept of innateness.

This concern is reasonable, and it is difficult to exclude the possibility that scientists do have an effective concept of innateness distinct from the vernacular concept but just failed to use it in our experiment. That said, we are skeptical for the following reasons. While the experiment is artificial and obviously different from the research done by our participants, it is not unrelated. All the cases describe real situations; they are not fictional; and the terminology we used is technical. Further, their topic—birdsong—is also commonly discussed in the debates about innateness in developmental psychology (e.g., Gallistel et al., 1991), comparative psychology (e.g., Marler and Slabbekoorn, 2004), and linguistics (e.g., Bolhuis, Okanoya, and Scharff, 2010). So, scientists are somewhat accustomed to thinking about the innateness of birdsong in a scientific context, and one would expect them to draw upon their distinctively scientific concepts of innateness to do so, if they had any.

Let's suppose now that we are right that scientists have not developed scientifically sound technical concepts of innateness, as some philosophers have argued, but rather rely on the vernacular concept of innateness. What is the significance of this fact? As we outlined earlier, the crucial point is that this concept results in inferences that are likely to lead from a true premise to a false conclusion. No trait is fully fixed (Griffiths and Machery, 2008), but the development of some traits is buffered against some specific environmental variation. People who deploy the concept of innateness in their thought would then infer from the fact that a trait seems fixed (i.e., fixed in some respect or other) that it is innate and, from the fact that it is innate, that it is present in most conspecifics (i.e., typical). But this inference is unreliable because these two dimensions, fixity and typicality, are not tightly connected empirically.

Inferring that a trait is typical because it is fixed in some respects will often lead to mistakes. The solution, some philosophers have proposed, is to eliminate the folk notion of innateness (e.g., Griffiths, 2002). But now if scientists do in fact use the vernacular concept of innateness rather than distinct, sound concepts of innateness, then the use of the concept of innateness should also be eliminated from science.

Perhaps a critic will respond that instead of eliminating the concept of innateness, philosophers and scientists should develop a sound concept of innateness; currently, scientists do not use scientific concepts of innateness, but they surely could, and developing such concepts is what should be done. We do not want to dismiss this argument too promptly, since some of us have said similar things for the concept of human nature (Griffiths, 2012; Stotz and Griffiths, 2018; Machery, 2017a, 2018), but we believe that the attraction which the vernacular concept of innateness seems to have on scientific thinking should lead us in another direction: better eliminate than develop new concepts and then fall back on the vernacular concept.

Conclusion

We have reported an exploratory study of the concept of innateness among scientists. Drawing on our past work on the vernacular concept, we have provided evidence that scientists' effective concept of innateness (in contrast to their explicit concept) is invariant across disciplines and is similar to the vernacular concept of innateness. This finding suggests that, at least in the case of innateness, the integration of a folk concept into a scientific conceptual toolkit does not lead to the formation of several distinct, scientifically sound concepts of innateness that fit particular theoretical and experimental niches; rather, the vernacular concept constrains the way scientists think about innateness. This finding reinforces the argument repeatedly made in recent decades that the concept of innateness should be eliminated.

Notes

1 By *concept* we simply mean the cognitive structure that is used in categorization judgments and that is somehow associated with the relevant word. So, the concept of innateness is the cognitive structure that is used in the judgments that something

is or is not innate and that is somehow associated with the word *innate*. There is no need for present purposes to specify the notion of concept more precisely, but one way to do so would be to follow Machery, 2009 and 2017b, Chapter 7. We will also at times refer to what scientists mean by a given word or to scientists' understanding of something to refer to the same cognitive structure.
2 We will usually drop *effective* in what follows. When we refer to scientists' concept of innateness, we have in mind their effective concept.
3 We are not committed to the view that the concept of innateness is a prototype, just that it represents prototypical features.
4 Because only a few individuals were classified in this post hoc manner, nothing substantial hangs on these decisions.

References

Atran, S. (1990). *Cognitive Foundations of Natural History: Towards an Anthropology of Science*. Cambridge: Cambridge University Press.
Baker, M. C. (2001). *The Atoms of Language: The Mind's Hidden Rules of Grammar*. New York: Basic Books.
Benedict, L. and Bowie, R. C. (2009). Macrogeographical variation in the song of a widely distributed African warbler. *Biology Letters*, rsbl20090244.
Bolhuis, J. J., Okanoya, K., and Scharff, C. (2010). Twitter evolution: Converging mechanisms in birdsong and human speech. *Nature Reviews Neuroscience*, 11: 747.
Boyd, R. and Richerson, P. J. (2005). *The Origin and Evolution of Cultures*. New York: Oxford University Press.
Carey, S. (2009). *The Origin of Concepts*. New York: Oxford University Press.
Claidière, N. and Sperber, D. (2007). The role of attraction in cultural evolution. *Journal of Cognition and Culture*, 7: 89–111.
Cheung, B. Y., Dar-Nimrod, I., and Gonsalkorale, K. (2014). Am I my genes? Perceived genetic etiology, intrapersonal processes, and health. *Social and Personality Psychology Compass*, 8: 626–37.
Dar-Nimrod, I. and Heine, S. J. (2011). Genetic essentialism: On the deceptive determinism of DNA. *Psychological Bulletin*, 137: 800–18.
Fodor, J. A. (1981). *Representations: Philosophical Essays on the Foundations of Cognitive Science*. Cambridge: MIT Press.
Gallistel, C. R., Brown, A. L., Carey, S., Gelman, R., and Keil, F. C. (1991). Lessons from animal learning for the study of cognitive development. In S. Carey and R. Gelman (Eds.), *The Epigenesis of Mind: Essays on Biology and Cognition* (pp. 3–36). Hillsdale, NJ: Psychology Press.
Gelman, S. A. (2004). Psychological essentialism in children. *Trends in Cognitive Sciences*, 8: 404–09.

Gelman, S. A., Ware, E. A., and Kleinberg, F. (2010). Effects of generic language on category content and structure. *Cognitive Psychology*, 61: 273–301.

Griffiths, P. E. (2002). What is innateness? *The Monist*, 85: 70–85.

Griffiths, P. E. (2012). Reconstructing human nature. *Arts: The Journal of the Sydney University Arts Association*, 31: 30–57.

Griffiths, P. E. and Machery, E. (2008). Innateness, canalisation and "biologicizing the Mind". *Philosophical Psychology*, 21: 397–414.

Griffiths, P. E. and Stotz, K. (2008). Experimental philosophy of science. *Philosophy Compass*, 3: 507–21.

Griffiths, P. E. and Stotz, K. (2013). *Genetics and Philosophy: An Introduction*. New York: Cambridge University Press.

Griffiths, P. E., Machery, E., and Linquist, S. (2009). The vernacular concept of innateness. *Mind & Language*, 24: 605–30.

Gross, S. and Rey, G. (2012). Innateness. In E. Margolis, R. Samuels, and S. P. Stich (Eds.), *The Oxford Handbook of Philosophy of Cognitive Science*. Oxford: Oxford University Press.

Haslam, N. and Ernst, D. (2002). Essentialist beliefs about mental disorders. *Journal of Social and Clinical Psychology* 21: 628–44.

Haslam, N. (2011). Genetic essentialism, neuroessentialism, and stigma: Commentary on Dar-Nimrod and Heine (2011). *Psychological Bulletin* 137: 819–24.

Hinde, R. A. (1968). Dichotomies in the study of development. In J. M. Thoday and A. S. Parkes (Eds.), *Genetic and Environmental Influences on Behaviour* (pp. 3–14). New York: Plenum.

Hull, D. L. (1965). The effects of essentialism on taxonomy: 2,000 years of stasis. *British Journal for the Philosophy of Science*, 15: 314–26.

Hull, D. L. (1988). *Science as a Process: An Evolutionary Account of the Social and Conceptual Development of Science*. Chicago: University of Chicago Press.

Hyundeuk, C. and Machery, E. (2016). Scientific concepts. In P. W. Humphreys (Ed.), *Oxford Handbook on the Philosophy of Science* (pp. 506–23). Oxford: Oxford University Press.

Kelemen, D. and Rosset, E. (2009). The human function compunction: Teleological explanation in adults. *Cognition*, 111: 138–43.

Knobe, J. and Samuels, R. (2013). Thinking like a scientist: Innateness as a case study. *Cognition*, 126(1): 72–86.

Lehrman, D. S. (1953). Critique of Konrad Lorenz's theory of instinctive behavior. *Quarterly Review of Biology*, 28: 337–63.

Linquist, S., Machery, E., Griffiths, P. E., and Stotz, K. (2011). Exploring the folkbiological conception of human nature. *Philosophical Transactions of the Royal Society B*, 366: 444–53.

Lynch, K. E., Morandini, J. S., Dar-Nimrod, I., and Griffiths, P. E. (2019). Causal reasoning about human behavior genetics: Synthesis and future directions. *Behavior Genetics*, 49: 221-234.

Machery, E. (2009). *Doing without Concepts*. New York: Oxford University Press.
Machery, E. (2017a). Human nature. In D. Livingstone Smith (Ed.), *How Biology Shapes Philosophy: New Foundations for Naturalism* (pp. 204–26). Cambridge: Cambridge University Press.
Machery, E. (2017b). *Philosophy within Its Proper Bounds*. Oxford: Oxford University Press.
Machery, E. (2018). Doubling down on the nomological account of human nature. In E. Hannon and T. Lewens (Eds.), *Why We Disagree about Human Nature* (pp. 18–39). Oxford: Oxford University Press.
Mameli, M. and Bateson, P. (2006). Innateness and the sciences. *Biology and Philosophy*, 22: 155–88.
Marler, P. R. and Slabbekoorn, H. (2004). *Nature's Music: The Science of Birdsong*. San Diego: Elsevier.
Medin, D. L. and Atran, S. (Eds.) (1999). *Folkbiology*. Cambridge: MIT Press.
Prinz, J. J. (2012). *Beyond Human Nature*. New York: W. W. Norton & Company.
Shtulman, A. (2017). *Scienceblind: Why Our Intuitive Theories about the World Are So Often Wrong*. New York: Basic Books.
Sperber, D. (1996). *Explaining Culture: A Naturalistic Approach*. Cambridge: MIT Press.
Stotz, K. (2008). The ingredients for a postgenomic synthesis of nature and nurture. *Philosophical Psychology*, 21: 359–81.
Stotz, K. and Griffiths, P. (2004). Genes: Philosophical analyses put to the test. *History and Philosophy of the Life Sciences*, 26: 5–28.
Stotz, K. and Griffiths, P. E. (2018). A developmental systems account of human nature. In T. Lewens and E. Hannon (Eds.), *Why We Disagree about Human Nature* (pp. 58–75). Oxford: Oxford University Press.
Warner, R. R. and Swearer, S. E. (1991). Social control of sex change in the bluehead wrasse, Thalassoma bifasciatum (Pisces: Labridae). *The Biological Bulletin*, 181: 199–204.
West, M. J., King, A. P., and Duff, M. A. (1990). Communicating about communicating: When innate is not enough. *Developmental Psychobiology*, 23: 585–98.
Zala, S. M. and Penn, D. J. (2004). Abnormal behaviours induced by chemical pollution: A review of the evidence and new challenges. *Animal Behaviour*, 68: 649–64.

Part Four

General Considerations

Causal Judgment: What Can Philosophy Learn from Experiment? What Can it Contribute to Experiment?[1]

James Woodward

Introduction

The topic of this volume is experimental approaches to philosophy of science. It is uncontroversial that empirical research, broadly speaking, is relevant to many topics in philosophy of science—philosophers interested in space and time should make use of the best empirical theorizing on these topics, and similarly for philosophers of biology interested in the structure of evolutionary theory, philosophers of psychology interested in perceptual processing, and so on. Issues having to do with the relevance of empirical, or more specifically experimental, research become more controversial, however, when we move to topics having to do with scientific method, or more generally, to issues that are within the purview of general philosophy of science—issues having to do with theory testing, confirmation, evidence, causal reasoning, explanation, and so on. A natural thought (and one I endorse here) is that these issues have an important *normative* component: they have to do with how we *ought* to think and reason. If so, one might then wonder how discoveries about how people *in fact* reason could be relevant to these normative concerns.

My aim in this essay is to address this issue in a particular context: causal reasoning. But I also have some broader aims. I will also compare appeals to experimental results with two alternatives: the traditional philosophical method of appealing to "intuitions" about cases, and more recent developments in "experimental philosophy," some (but by no means all) of which appeal to survey-like results to address philosophical questions. My overall goal is to use this work to explore some more general questions having to do with how empirical results might best be brought to bear on issues in philosophy of science. Causation and causal explanation (hereafter, I will sometimes just write "causation") are of course topics of long-standing importance in philosophy of science and in philosophy more generally. These topics have also recently undergone a rich theoretical development in disciplines like statistics, econometrics, and computer science. At the same time, the past several decades have seen an explosion of research on the empirical psychology of causal and explanatory reasoning, much of it conducted by psychologists, but also involving contributions from researchers in philosophy, primatology, and animal behavior, among other disciplines. It is natural to suppose that these bodies of work can usefully inform one another, but the details of how this might work are far from obvious.

Insofar as my focus is empirical, it will be on features of causal reasoning (such as the role of invariance) that are reflected directly in scientific practice. Much of the work having this character has been conducted by psychologists such as Gopnik, Cheng, and Tenenbaum. By comparison, there has been much less work on causal judgment and cognition by experimental philosophers, and the research that has been conducted (e.g., Hitchcock and Knobe, 2009) has tended to focus on actual cause judgments and on the role of norms in causal selection. Although important for understanding aspects of lay causal thinking, these issues are (in many cases) less central to causal thinking in science, which tends to focus on type-causal claims.[2] Hence my focus on empirical research conducted by psychologists.

Normative and descriptive theorizing

Normative theories of causal reasoning purport to describe how we ought to think about causation or what good or correct causal reasoning consists in; descriptive theories describe how we do reason and the causal factors (including the computations and information processing) that underlie such reasoning. A normative focus is explicit in work on causation in disciplines like statistics and machine learning. It is also the dominant way of thinking about theories of causation in empirical psychology that have descriptive aspirations—a point that is less paradoxical than it may initially seem, for reasons to be explained. It is less common to view the standard philosophical theories of causation as normative proposals, but I suggest this is a very natural and illuminating way of understanding their significance. Here the normative content has less directly to do with inference but instead with how we should conceptualize causation (including distinctions we ought to draw among various kinds of causal claims), which relationships we should treat as causal, and how causal claims should be connected to other notions of interest, such as counterfactual dependence and probability. For example, David Lewis' well-known theory might be regarded as a normative proposal to the effect that we should regard only those events that stand in a certain relationship of counterfactual dependence as causally related, that we should reason about causal relationships by means of such counterfactuals, assessing these in turn in terms of the particular similarity measure across possible worlds advocated by Lewis and so on. Theories according to which causal relationships must involve the transference of energy and momentum (so that, e.g., absences cannot be causes) can be regarded as proposals that we ought to regard certain relationships and not others as causal.

The characterization of the above theories as "normative" raises the obvious question of where their normativity comes from. I address this issue in more detail in the section titled "Sources of normativity," where I argue that this is a matter of means/ends justification: there are certain ends or goals distinctively associated causal reasoning, and we justify claims that this reasoning should possess various features by showing that these features conduce effectively to these goals.[3]

Thinking about philosophical theories of causation as normative in this way contrasts with two more standard ways of thinking about their significance, according to which they are either attempts to capture the content of "our concept" of causation or else (for those who are more metaphysically inclined) attempts to describe the "nature" of causation or "causation as it is in itself." From my perspective, such construals efface the normative content of these theories and make it sound as though their intent is purely descriptive or reportorial (of our concept of causation or causation itself).[4] An additional problem is that it is dubious that there is any such thing as "our concept" of causation if by this is meant a single "concept" shared by all competent users. Instead, even from the point of view of a single overarching treatment of causation, such as a counterfactual or interventionist theory, there are a number of distinct varieties of causal relationship that should be distinguished: actual cause, various type-causal notions, and notions of direct and indirect effect, as well as different notions of indirect effect.[5] In addition to this, empirical research (some of which will be described below) shows that people exhibit nontrivial heterogeneity in some causal judgments. If such judgments are interpreted as evidence for some concept of causation that "we" possess, this raises the question of whether those who judge differently have different concepts. In interpreting theories of causation as claims about concepts of causation, we saddle ourselves with the problem of distinguishing those features of practices of causal reasoning and judgment that are reflective of or constitutive of the concept (or concepts) of causation from those that are not. Moreover, on any reasonable account of concepts, there will be many interesting and important features of causal reasoning that are not constitutive of our concept of causation—examples will be provided below. In fact, experimental results in the cognitive psychology of causal reasoning are rarely presented as claims about the (or a) concept or concepts of causation—instead, researchers talk about different sorts of representations of causal relations and the computations associated with these, different strategies for causal learning and judgment, and so on. Philosophers would do well to follow this practice, whether their goals are descriptive or normative.[6]

The other standard construal associated with philosophical theories is that these aim to characterize "what causation is" or to capture its underlying "nature," as distinct from our concept of it. A sharp version of this distinction (between concept and underlying nature) only makes sense if how we think about or conceptualize causation can come apart very sharply from what the causal relation "really is," with the latter being revealed by some science (e.g., physics) distinct from common-sense reasoning or perhaps revealed by some form of metaphysical insight. (An analogy would be the distinction between how we think about gold, as reflected in its stereotype as a valuable, yellowish metal and what gold really is—an element with atomic number 79.) Whatever one thinks of this possibility of uncovering the underlying nature of causation, it involves a project to which the cognitive psychology of causal reasoning as well as the methods of consulting intuitions about particular cases which are common in the philosophical literature appear to be largely irrelevant (or if relevant at all, only in a negative way)[7]—instead, it will be physics or metaphysics that tell us what causation really is. I will accordingly ignore this project in what follows.[8]

Thinking about philosophical theories of causation as (in part) normative proposals fits naturally with another important idea which is commonly assumed in the psychological literature and which I think ought to guide philosophical inquiry as well. This is that human causal reasoning is often very successful in enabling us to cope with and get around in the world. Different theories of causation can be associated with different views about the coping abilities that causal reasoning provides—as discussed in more detail below, for interventionists (cf. Woodward, 2003) these distinctively have to do with manipulation and control.[9] As an empirical matter, our practices of causal cognition are relatively "rational" or well adapted to the circumstances in which we find ourselves—something that becomes particularly salient when we compare the abilities of humans (even young children) with other primates. (See, for example, Woodward, 2007.) Just as one of the goals of a theory of the operation of the human visual system should be to understand how it is that this system can successfully extract from the visual array reliable enough information about the structure of the external environment, so also theories of causation and causal reasoning should give us some insight into how we are able to successfully learn about and reason concerning the causal structure of our environment. As I note below, this is one reason (among many) why the most successful descriptive investigations into causal cognition are often those in which some normative theory plays an important role—one of the things we want from a good descriptive theory is to explain how we are successful to the extent that we are, and this requires normative theorizing (as well as a conception of what success consists in). This focus on explaining success provides one of a number of points at which the normative and the descriptive elements in theorizing can fruitfully come together.

An emphasis on explaining success and well-adaptedness in connection with causal cognition also points to an additional limitation of the idea that the goal of theories of causation should just be to characterize our concept of causation: even putting aside the misgivings expressed earlier, what we want to understand is not just what concepts we have but also why those concepts work or lead to successful reasoning and inference to the extent that they do and, along with this, what their limitations might be. When limitations are present, we also ought to consider how our current ways of reasoning about causation (including our "concepts") might be improved so that they are better adapted to our goals, an enterprise that may involve rethinking or re-engineering those ways of reasoning. By itself, a description of our current concepts does not accomplish this.

I intend these claims about the adaptiveness or rationality of much commonsense causal reasoning as a high-level empirical claim—one that I think is supported by a great deal of current psychological research (see, e.g., Holyoak and Cheng, 2011). A detailed defense of this claim is beyond the scope of this essay, but some brief clarificatory remarks may be helpful. First, what about various well-known results that seem to show that humans are prone to all sorts of errors when reasoning about probability and other matters? Some of these results seem to me to be infected with the very same methodological problems that often infect philosophical accounts that trade heavily on appeals to intuitions as well as survey-style experimental philosophy (X-phi) research—for example, failures to control for ambiguities in the question the philosopher (or psychologist) poses to herself or others, with the result that subjects look irrational when they are simply answering a different question from what the researcher intends. Some of these methodological problems will be discussed below Second, there is a crucial distinction between people's explicit reasoning capacities about probabilities and other matters when presented with verbally described problems, and their abilities to make use of probabilistic and other sorts of information when this is presented in formats and contexts which do not require such explicit reasoning. To a very substantial extent, people are better at tasks of the latter sort than the former. Many nonhuman animals as well as humans are, as an empirical matter, very good at tracking frequency and contingency information when this is presented in an ecologically natural way and adjusting their behavior in the light of this in ways that are rational, given their goals. Brains appear to do lots of Bayesian updating, even though many humans are unable to successfully reason explicitly with Bayes' theorem. Focusing on people's explicit reasoning abilities in responding to verbal questions underestimates the rationality of much that they do.[10]

Finally, let me make explicit another assumption that goes along with these claims about adaptiveness. I focus in this essay on aspects of causal cognition among ordinary, lay subjects. I think, however, (this is another empirical claim) that there is a great deal of continuity between such ordinary causal cognition when successful and more sophisticated forms of causal reasoning found in the sciences.[11] Consider the assumptions about the invariance of causal relationships that underlie Cheng's causal power theory, discussed in the section "Invariance: Normative theory and descriptive results." Very similar assumptions play an important role in causal and explanatory reasoning in many areas of science (Woodward, 2010, 2018). This is one of several reasons why we can learn things that are relevant to science and philosophy of science by studying common-sense causal reasoning.

Sources of normativity

I've been talking so far of philosophical and other theories of causation as normative proposals. What is the source of this normativity, and how might we assess whether one proposal is normatively superior to another? In this essay, I will treat this as entirely a matter of means/ends justification. Inquirers have certain ends or goals, including epistemic goals—achievement of these goals is what constitutes success. Proposals about causation—whether these are proposals about how to infer to causal conclusions from certain kinds of data or which patterns of causal reasoning are correct, or how we should conceptualize or think about causation or make distinctions among different sorts of causal relationships—are to be evaluated in terms of whether or not they are effective means to these goals. In principle, this approach to justification might be associated with a number of different goals. As noted above, interventionists think that one of the distinctive goals associated with causal thinking is the discovery of relationships that are exploitable for purposes of manipulation and control, but (as far as this approach to justification goes) one might instead associate causal reasoning with other sorts of goals—for example, with the simple and nonredundant representation of information about regularities. In any case, various causal concepts and strategies for causal inference are to be evaluated in terms of how well they conduce to this goal. For example, a number of arguments show that randomized experiments are a particularly good way of identifying relationships that support manipulation. Possible procedures for inferring causal conclusions from nonexperimental data can then be evaluated in terms of the extent to which they yield information that is like the manipulation—supporting information that would result from a properly conducted experiment; this is the generally accepted rationale for employing techniques like instrumental variables and regression discontinuity designs (see, e.g., Angrist and Pischke, 2009). This amounts to a means/ends justification for the adoption of these techniques.

As another illustration, given our general interests in identifying relationships relevant to manipulation and control, we have a more specific interest in formulating manipulation-supporting relationships that generalize successfully to contexts that are different from those in which they were originally discovered. We thus value the discovery of causal relationships that are relatively invariant (where this means, among other considerations, that they will continue to hold under changes in background circumstances—this will be discussed below) and it makes sense that we adopt ways of thinking about causation that reflect invariance-linked features, such as Cheng's causal power concept (see "Invariance: Normative theory and descriptive results" and "Cheng's causal power model").[12]

Other illustrations of the same basic means/ends justificatory strategy, not necessarily connected to causal inference, are provided by classical statistics. For example, the choice of an estimator for some quantity of interest proceeds by postulating certain goals or criteria that the estimator should satisfy—for example, that it should be unbiased and (among the class of unbiased estimators) have minimum variance. Justifying the choice of an estimator is then just a matter of identifying the estimator that best achieves these goals. As this example illustrates, showing that some concept or set of features is well adapted to some goal often involves mathematical or conceptual analysis, although it may also have an empirical component. Again, there is a relatively close analogy with understanding the visual system, where this is taken to involve ascription of goals to that system (accurate enough representation of aspects of the environment relevant to action) and then an investigation (usually with a substantial mathematical component) of the means the system employs to achieve these goals.[13]

Relating the normative and descriptive

The framework in the sections "Normative and descriptive theorizing" and "Sources of normativity" yields several ways in which descriptive and normative considerations can be related in studies of causal reasoning. One possibility is that we find that, as a matter of empirical fact, causal cognition among humans (or other subjects) exhibits feature F where F may be, for example, the use of a causal concept with a certain structure or certain inferential or reasoning strategies involving causation. We can then ask whether feature F contributes to some goal G associated with causal reasoning (for interventionists, a goal such as manipulation or some subsidiary goal that follows from this). If so, we then have a partial explanation of those subjects' "success" in their causal cognition to the extent that success is characterized in terms of the achievement of G: subjects succeed because their causal cognition exhibits feature F. (The examples below involving the role of invariance considerations in causal reasoning have this character.) Note that in this reasoning, a feature that is present as a descriptive matter is linked to a feature that involves a normative characterization, thus providing an is/ought connection. An additional possibility is that we may also have grounds for believing that some sort of selective process is at work that *explains* the presence of F in the sense that F is selected for *because* it leads to G— thus accounting for the presence of F via a functional explanation. Such selective processes might involve either natural selection or learning—for example, subjects learn a certain way of thinking about causal relationships (rather than some alternative) through some feedback process involving the satisfaction of their goals with the feedback reinforcing that way of thinking.

In following either of the approaches described above, we identify features F of causal reasoning through empirical investigation and then ask, concerning them, whether they have a normative rationale in terms of contributing to goal G. I emphasize that the features F are *not* regarded as normatively justified or appropriate merely because they are found in people's causal reasoning. Instead, to the extent that they are normatively justified, this is because they can be given an independent justification in terms of contributing to G. I will add, though, that the empirical discovery that causal reasoning exhibits feature F can prompt us to consider the possibility that F may have some normative justification, where we might not think to consider this possibility prior to the discovery of F. For example, the tendency of subjects to assign higher causal strength ratings to contingencies in which the base rate of the effect is high, commonly thought to lack a rational justification, can be given a normative justification in terms of the invariance-linked ideas developed in Cheng's causal power theory, as explained in the section "Cheng's causal power model".

Moreover, the direction of discovery can go, so to speak, in the opposite direction: given a normative theory which tells us that normatively good causal reasoning will contain feature F, we can then empirically investigate the reasoning of adult humans and other subjects to see if they exhibit this feature. In a number of cases, the answer to this question turns out to be "yes."[14] Such cases illustrate what I will call the *motivating* or *enabling* role of normative theory in connection with empirical investigation; often it does not occur to researchers to do certain experiments or to look for whether certain features are present in causal cognition in the absence of a normative theory that tells us that normatively appropriate cognition will exhibit these features. This is another way in which normative and descriptive investigations can fruitfully interact.[15]

Intuition and X-phi

The picture just sketched contrasts both with the usual forms of the intuition-based approaches to causation common in portions of the philosophical literature and some approaches that are common in X-phi. I begin with the former, which I see as proceeding as follows: the philosopher describes various cases and then consults his or her "intuitions" about these, with the operative question usually being whether these are instances of causation. For example, a case in which a gardener fails to water flowers and the flowers die will be described and then an intuition or judgment will be advanced about whether the gardener's omission caused the flowers to die, and similarly for cases involving double prevention, causal pre-emption, and so on. The task of constructing a normative theory of causation is then taken to be, at least in part, the task of constructing a theory which captures or reproduces these intuitions—that is, one part of the test for whether one has produced a good theory is whether it agrees with intuition. An obvious initial problem is that the intuitions of any single person are unlikely to be fully consistent with one another and, moreover, different people may hold inconsistent intuitions. In the absence of a single theory that fully captures everyone's intuitions, a standard response is to appeal to some version of reflective equilibrium— one looks for a theory that captures as many intuitions as possible, but where this may involve rejecting some intuitions in favor of others in order to maximize overall systematic coherence.[16] Additional constraints/desiderata that are motivated on more general philosophical grounds such as the demand that the resulting theory be "reductive" or that it reflect certain requirements allegedly coming from "fundamental physics" or from "metaphysics" may be added to this mix—a prominent recent example is Paul and Hall, 2013, and a similar program seems to underlie Lewis' well-known work on causation.

A general problem for any approach that takes normative justification to involve appeal to reflective equilibrium is that there is no reason to suppose that there is a unique outcome which represents the best possible trade-off among the different intuitions and other desiderata (and no obvious way of telling whether we have found such an outcome, supposing that it exists). Instead, there may be multiple equilibria, each corresponding to different ways of trading off or balancing the desiderata just described; or there may be no equilibrium. Indeed, the variety of different theories that have been produced by investigators claiming to follow something like (some version of) the method of reflective equilibrium seems to support the conclusion that either there are multiple equilibria or that all but one of the theorists have misapplied the procedure (and we can't tell which). The approach to normative justification that I favor does not suffer from these difficulties, since it takes the basis for such justification to involve means/ends reasoning rather than an appeal to reflective equilibrium.[17]

But even putting this aside, there are several additional difficulties. One is that it is unclear what the goal or point of the enterprise just described is. It can't be intended just as empirical psychology or description/explanation of aspects of causal judgment, since even if we accept that intuition is a source of information about these (on this see immediately below), the project under discussion involves rejecting some of these intuitions in favor of others, settling cases in which intuition is uncertain in definite ways (rather than just reporting the uncertainty), and subjecting the whole investigation to additional philosophical or physics-based constraints that are not motivated by empirical psychology. A straightforwardly descriptive enterprise (with intuitions just taken to be reports or descriptions of how people judge) would not take this form. But at the same time, the account that will emerge from such a procedure will not have an obvious functional or normative rationale (in the sense described in "Normative and descriptive theorizing" and "Sources of normativity") either. One way of seeing this is simply to note that the reflective equilibrium procedure as described assigns no role to what the end product is to be used for or whether it is well or poorly designed to achieve goals associated with causal thinking. Instead, the product looks like a curious hybrid, partly constrained by the goal of capturing intuitions and partly constrained by other sorts of considerations coming from philosophy or the metaphysics of science, which seem to have little to do with functional/normative considerations. The alternative approach to normative theory and its connection to empirical psychology described in "Normative and descriptive theorizing" does not suffer from these limitations.

So far, I have not addressed the question of what "intuition" itself can tell us (its "reliability" or what kind of information, if any, it provides) and how this consideration impacts assessment of traditional intuition-driven approaches. (Recent discussion of the role of intuition in philosophy has tended to focus on this question.) One possibility is that intuition involves a kind of rationalistic grasp of facts about the subject of the intuition—our intuitions about causation provide us with some sort of purely reason-based acquaintance with the nature of causation, and so on. It is hard to construct a plausible story about how such a "faculty" could be a reliable source of information, and I will not consider this possibility further.

Another, prima facie more plausible, possibility, defended by Goldman (2007) among others, takes claims about intuitions to reflect (at least in the right circumstances) claims about concepts that ordinary speakers acquire as part of their linguistic competence. Thus intuitions about examples involving causation reflect the speaker's knowledge of how causal concepts are correctly applied and hence information about the structure of those concepts. One limitation of this view is that, for reasons described previously, it assigns too central a role to concept talk in causal cognition, and in effect takes causal cognition to be just a matter of applying causal concepts to scenarios, thus neglecting many other aspects of causal cognition having to do with causal learning and inference. However, we might broaden Goldman's suggestion in the following way: when a philosopher reports the intuition that such and such is a case of C (e.g., causation), that person at least provides evidence about how she judges. If, as is very often the case, she believes or assumes that her intuition is fairly widely shared (as might be indicated by her use of words like "we think" or "we would say") and if there is reason to assume that this assumption is accurate, then the report of the intuition can in principle provide us with evidence about shared practices of judgment . To the extent such claims are correct (a nontrivial condition of course), we can use them just as we use other sorts of empirical evidence, employing normative theory to assess their appropriateness in the manner described in "Normative and descriptive theorizing."[18] Moreover, claims about the extent to which such judgments are widely shared can in principle be checked empirically, by survey-type methods that are commonly employed in X-phi.

Of course, the content of any particular intuitive judgment that, for example, the relationship in some scenario involving double prevention is causal will be "about" that scenario and the nature of the relationship described in it. However, I assume, in virtue of earlier arguments, that we cannot understand the mere having of such judgments as evidence for the truth or correctness of what is asserted in the judgment (e.g., as evidence that the relationship in question is "really causal" (whatever that might mean) or as evidence about the "nature" of causation). So if the intuitive judgment is evidence for anything, it must be evidence for something else. Goldman takes the judgments to be evidence about the structure of our concepts (even though the content of the judgments does not directly have to do with our concepts); I have suggested broadening this to take the judgments to be (in some cases) evidence for or as implying claims about shared practices. In other words, I suggest that insofar as appeals to intuition or judgments about cases have a *legitimate* evidentiary role, it is this.[19]

If we think of intuitions in this way, we have a defense of their role in philosophical and other sorts of argument—a defense that is thoroughly naturalistic in spirit. This is because information about shared practices can, as explained above, play a legitimate role in philosophical argument, especially when combined with other assumptions. However, this defense is limited in important respects—some obvious, some less so. One limitation, already noted, is that to the extent that intuitions just report judgments about cases, they are unlikely to be a useful source of information about other aspects of causal cognition, such as learning. In addition, such judgments often do not provide reliable information about the causal processes that underlie them (See "Intuitions, surveys, and causal inference"). Another limitation, which intuitions share with the use of verbal responses to verbal probes more generally (whether these are responses to surveys or judgments in experimental situations as is the case with a number of the experiments discussed below), is that the question the intuiter poses to herself may be unclear or ambiguous in ways that are not appreciated, either by the intutier or her audience (See "Cheng's causal power model").

It should be clear that this limited defense of appeals to intuition does not support a number of the uses to which such appeals have been put in the philosophical literature. First, note that on this conception, the mere having of an intuition, even if this is widely shared, does *not* by itself have implications for what the normatively best account of causation or causal reasoning is (or, as I have argued, what the nature of causation is). What appeals to intuition show (at most) is that people make certain judgments—as argued above, whether these judgments are normatively appropriate requires appeal to an independently grounded normative theory.[20] This stands in contrast to the tendency of many philosophers to think of intuitions about cases as prima facie self-warranting in the sense that the intuition itself provides evidence of the normative correctness (or "truth") of what is reported in the intuition. Examples discussed below suggest that this picture of the role of intuition is naïve: for example, a number of subjects report judgments of causal strength regarding certain scenarios that are arguably normatively mistaken, although an approximately equal number also report different judgments that are normatively correct. If we think of these judgments as reports of intuitions, it is clear that both sets of intuitions cannot be correct and that some external standard for correctness is required. By the same token, however, we also see that the fact that an intuition or judgment is not universally held does not by itself show that it is "wrong"—another point to which I will return.

Next, I turn to yet another underappreciated limitation on appeals to intuition about cases. This is that although such appeals may in principle provide information about what people's judgments *are*, they usually (or at least often) are *not* reliable sources of information about *why* people judge as they do or about the factors on which their judgments depend or about the representations and computations that underlie these judgments.

Intuitions, surveys, and causal inference

To develop these points, I want to introduce some additional analytical distinctions. I begin, however, with some remarks about what I am *not* claiming. I distinguish below between bare surveys, which record what judgments people make, and causal analysis, which attempts to uncover the factors that cause their judgments and the psychological processes that underlie these. A bare survey, however well conducted, does not by itself provide a causal analysis. This does not mean that surveys are unimportant or uninteresting—on the contrary.[21] I take some of the research conducted in X-phi to be survey-like—I will give examples below. Moreover, as already intimated, many traditional armchair appeals to "intuition" can be thought of as like surveys in some relevant respects, but with a very small N. However, I do not claim—and it is not true—that all research in X-phi is survey-like; some significant portion of it aims at and succeeds in providing causal analysis—again I will give examples below.

Bare surveys

Suppose that a researcher exposes a group of subjects to a scenario or case description or a group of these and then asks for judgments about them. For example, subjects might be presented with a scenario in which a gardener fails to water plants and the plants die, with the subjects being asked whether the gardener's omission caused these deaths; or subjects might be presented with a Gettier-type scenario or a set of these and asked whether person described in the scenario "knows" that such and such is the case, as in some of the research reported in Turri (2016). If this is all that is done, I will call this procedure and the results it produces a *bare survey*. Of course, a bare survey can be well or badly designed—examples of the latter occur when the questions employed in the survey are unclear or confusing, when subject responses are influenced by their expectations about what the experimenters are looking for, when there are order effects, or when the subjects to whom the survey is given are unrepresentative of the population to which one wants to generalize. At least sometimes, such methodological problems can be adequately addressed by improving the survey design—surveys are not intrinsically flawed just because they are surveys. However, although when methodological problems are adequately addressed, a properly designed bare survey can provide information about what people in the population of interest judge, there are many other questions it cannot answer—in particular, a bare survey cannot tell us why subjects judge as they do (what causes them to judge as they do or what processing underlies their judgment—see "Causal analysis").

I suggested above that reports of intuitions are most charitably construed as implying claims about shared practices of judgment. In this respect, they resemble claims supported by bare surveys, although (of course) with the difference that, in the case of intuition, the response(s) may be those of a single person who is identical with the researcher or alternatively may be the responses of a small set of colleagues. This seems to fit many of the examples of reports of intuitive judgment in the causation literature. For example, Dowe (2001) considers a number of examples of alleged causation by absence and reports that these do not seem to him to be genuine cases of causation, or at least that they seem "different" than more standard cases of causation which he thinks involve transfer of energy and momentum. By contrast, Schaffer (2000) describes a number of examples of double prevention involving mechanical devices and reports that his judgments are that these are straightforward cases of causation. Lewis (1986) considers cases of symmetric overdetermination (as when two riflemen shoot a victim, with either shot being sufficient for death) and claims that common sense delivers no clear verdict about whether either of the individual shots causes the death.

We may think of each of these authors as conducting surveys of their own responses to cases, which they seem to expect their readers to share. To the extent that readers share their responses, this is information about what "people" or some substantial number of them think about the causal status of omissions, causation by disconnection, and symmetric overdetermination. Of course, similar (and perhaps more reliable) information might be obtained by means of a conventional survey of judgments about hypothetical scenarios conducted over, for example, Mechanical Turk. To the extent that we are interested in undermining or supporting claims about what most people would say or what the folk think, survey-style results can be very valuable.[22]

Surveys with covariation

In a somewhat more ambitious undertaking, the researcher might explore how the judgments of some group of subjects covary with some other variable. Unlike a bare survey (which requires recording only subject responses), this *requires* that some additional variable besides those responses be measured and that there be variation in both the responses and the additional variable. Speaking generically, we may distinguish (at least) two possibilities. First, different subjects (or the same subject on different occasions) may be presented with different scenarios (scenarios whose content varies along some dimension) and differences in subject judgments recorded with the aim of determining whether there is covariation between differences in the content of the scenarios and the judgments. For example, oversimplifying somewhat, Walsh and Sloman (2011) presented subjects with two scenarios, one of which involved a double prevention relation and the other a connecting process (a marble knocks over a domino). Subjects were asked whether the relationships in the scenarios were causal. A higher percentage judged that causation was present in the second scenario than in the first. Here there is covariation between the content of the scenarios and subjects' judgments.

A second possibility is to present subjects with a single scenario about which different subjects judge differently. The researcher then determines whether there is covariation between these judgments and other variables, which may include demographic factors. For example, Machery et al. (2004) presented "Western" and "Asian" subjects with scenarios such as Kripke's Godel case, asking them to make judgments about the referents of names. These authors claim that the Asian subjects were more inclined to make "descriptivist" judgments about reference than Westerners.

Such surveys with covariation can be valuable for a number of reasons. Most obviously, they can show that judgments that were assumed to be universal, or nearly so, vary considerably across subjects, including different demographic groups—thus casting doubt on claims about "widely shared intuitions." Moreover, to the extent that such judgments not only vary but covary with factors such as cultural background that we think are irrelevant to whether the judgments are "true" or "correct," this seems to further undermine a number of the standard uses to which traditional philosophers have attempted to put them. On the other hand, the mere observation of a covariation between subject responses and some other variable (even if the covariation is genuine in the sense that it holds in some target population to which one wants to generalize) does not by itself establish that the second variable causes the variation in responses. This is the goal of causal analysis.

Causal analysis

In causal analysis, the goal is not just to describe subject responses or judgments or the variation that may occur in these but also to discover the cause or causes of such variation and/or the processing or mediating variables that underlie it. I follow conventional ideas about causal inference in holding that in order to do this successfully, one must rule out the possibility that other factors besides the candidate cause are responsible for the variation in question—as it is often put, one must rule out or control for possible confounding factors. *Confounding* is present when there is covariation between subject responses R and some other factor X but this covariation does not arise because X causes R but rather is due to the operation of other factors—for example, a common cause of X and R. (Note that this is a different problem than the methodological problems that arise in connection with survey design such as nonrepresentativeness and experimenter effects.)

In principle, there are two different possible ways of controlling for confounders—experimentation and causal modeling. (These may also be used in combination.) The crucial feature of an *experiment*, as I shall use this term, is that there is active manipulation of the putative causal factor in a way that makes it independent of other possible causal factors that may influence the effect. When there is covariation between such an independently manipulated candidate cause and the effect, this is taken to show that the candidate cause is a genuine cause. In an experiment, such manipulation may be accomplished by randomization or by independent control of these alternative causal factors, assuming that they are known. As an illustration, Vasilveya et al. (2018) explored whether differences in the perceived background invariance or insensitivity of causal claims (the extent to which those claims continue to hold under variation in background circumstances) cause differences in judgments of causal strength involving those claims. Different causal claims can differ along many different dimensions, and to establish that differences in the insensitivity of these claims were responsible for differences in subject's judgments, Vasilveya et al. needed to rule out the possibility that such other differences were causally responsible for the difference in judgment. For example, they needed to rule out the possibility that differences in Δp = (the probability of the effect conditional on the presence of the cause minus the probability of the effect conditional on the absence of the cause) associated with the causal claims employed caused the difference in subject judgments. They accomplished this by using causal claims that were matched for Δp but differed in background invariance. Such careful thinking about what might be a confounding factor and taking steps to control for it is essential for reliable causal inference.

An alternative strategy for causal analysis involves causal modeling of the factors that influence subject responses, where this may, but need not, make use of experimental manipulation. Such modeling requires measurement of possible confounders and other variables and the application of some causal inference procedure (there are many candidates for these—constraint-based procedures such as those described in Spirtes et al. [2000], Bayesian analyses [e.g., Tenenbaum, Griffiths, and Kemp, 2006], and various structural equation methods, among others). Again, in all of these methods, the goal is to show that subject responses R are caused by some factor X by ruling out competing explanations for the covariation between X and R. Causal modeling can of course be carried out on purely "observational" data in which there is no experimental manipulation, so experimentation in the sense just described is not required for reliable causal analysis, although in empirical psychology, it is typical to rely at least in part on experiments.[23]

To illustrate the difference between successful causal analysis and a survey with covariation, let us return to the research conducted by Walsh and Sloman. These authors show (let us assume) that there is covariation between subject judgments of causation and whether the presented scenarios involved (i) a connecting causal process or (ii) dependence without such a process. Walsh and Sloman infer from this that this difference between (i) and (ii) is what caused this difference in causal judgment. In my view, they are not entitled to infer this on the basis of the data generated in their experiment. The problem is that they have not ruled out the possibility that some other difference between their two scenarios influences the difference in causal judgments.[24] For example, their disconnection scenario differed from the scenario involving a connecting process not just in terms of whether such a process was present but also in terms of the relative invariance of the dependence relations present in the two scenarios. If subject judgments are influenced by such differences in invariance (with less invariant relations judged as less causal or noncausal), it could be this difference which is responsible for the difference in judgment. In fact, Vasilyeva et al. (2018), as well as Lombrozo (2010), provide experimental evidence that this is the case. A similar analysis may apply to intuitions about causation by absence, as suggested in Woodward (2006). Dowe, for example, notes that he and others judge that some cases in which there is a relation of dependence between an absence and an outcome are noncausal and infers that he and others make this judgment *because* there is an absence of a connecting process in the examples considered. Even supposing (as seems plausible) that he is correct about how most people would judge regarding his examples, it does not follow that he is correct about what features of his scenarios cause people to make these judgments. To show this, one must control for other factors besides the absence of a connecting process that may drive judgments in causation by absence scenarios. Examples like this illustrate that even if one is a good judge of how others will judge regarding various scenarios, it is easy to be misled about why oneself and others judge as they do—introspection often is not a good guide to this, and more rigorous causal analysis is required instead. It is thus important that the psychological research described below in "Cheng's causal power model" involves genuine experiments that support causal analysis, not just surveys or surveys with covariation.

Despite this, it is fairly common to find philosophers not just reporting their intuitive responses (and claiming that others will have similar responses) but also either explicitly claiming that they can tell which factors are causally influencing those responses or at least writing as though they have reliable introspective access to information about this. Sometimes the argument goes like this: the philosopher finds two cases that (it is supposed) match exactly in all possibly relevant respects except for the presence or absence of a single feature X. The philosopher takes this to be a case in which she is controlling "in her mind" for all the other relevant differences between the two cases besides X— that is, the philosopher thinks of herself as running a controlled experiment, albeit in her mind. The philosopher finds that her intuitive judgments about the two cases differ and attributes this to the difference made by X. (See, for example, the discussion in Kamm [1993], who explicitly endorses this method.) The obvious problem with this procedure is that it requires that the philosopher has introspective access to all of the other factors besides X that might influence her differential response and also that she can recognize when one of these is present and influencing her response and somehow remove or correct for the factor in question. I don't mean to claim that people can *never* do this, but I'm dubious that philosophers or anyone else can reliably execute such mentalistic analogs to an actual experiment in many cases of philosophical interest and that, moreover, they can be in a position to *know* that they have successfully done this. Indeed, both in the literature on causal cognition and elsewhere, there are many experiments that show that what is actually influencing people's differential responses is not what they judge to be influencing them. To this we may add that if people really did have reliable introspective access to whatever is influencing their judgments, experimental psychological investigations into this would be unnecessary—everything could be done from the armchair.

Let me add that in distinguishing between surveys (with or without covariation) and causal analysis, I certainly do not mean to claim that experimental philosophers only do the former. In the field of causal cognition alone, examples of (in my judgment) convincing causal analysis carried out by experimental philosophers include Hitchcock and Knobe (2009), showing the influence of norms on causal judgment; Kominsky et al. (2015), showing factors influencing actual cause judgment and competition between causes; and Icard et al. (forthcoming), showing the influence of normality judgments on actual cause judgments.

In discussions of X-phi, both pro and con, there is a tendency to focus on the difference between traditional armchair methods and surveys of what the folk think, with some experimental philosophers arguing for the superiority of the latter and more traditional philosophers rejecting such claims. However significant this difference may be, there is also an important divide that puts surveys and appeals to intuitions on one side and causal analysis, whether carried out by experimental philosophers or by psychologists, on the other. In my view at least, sometimes a philosopher in the armchair may have a good sense about how others will judge, although of course it will always be an empirical issue to what extent this is the case. However, even if armchair methods are sometimes reliable in this application, it is a further question whether they can be used reliably in causal analysis—and here the answer seems to be negative, at least in many cases. If this is correct, the most innovative forms of X-phi are those that involve causal analysis—in contrast to surveys, they address questions for which armchair analysis seems particularly unsuited.

Invariance: Normative theory and descriptive results

I turn now to a discussion of some more specific psychological hypotheses about causal cognition and associated experimental results. I will try to show how these illustrate the general methodological ideas I have described. My focus will be on hypotheses and results having to do with the role of *invariance* in causal cognition.

The general idea that motivates this research has both a descriptive and a normative component. The descriptive component is that we tend to think and reason about causal relationships in terms of invariance, and that, other things being equal, we prefer, when we can discover them, causal relationships that are more invariant rather than less. The normative element is that it is correct or appropriate to reason in this way since relatively invariant relationships better satisfy goals associated with causal reasoning. The general idea of invariance is that a relationship $C \to E$ is more invariant to the extent that it would continue to hold as various other factors change—"continue to hold" means that the relationship continues to apply or to correctly describe what is going on. These "other factors" come in a variety of different forms, corresponding to different aspects of invariance. For example, we can ask whether the $C \to E$ relationship continues to hold if various other factors distinct from C and E, ("background factors"), change.[25] Another aspect of invariance has to do with whether the $C \to E$ relationship would continue to hold under changes in the values taken by C or changes in the frequency with which those values occur. The research described below makes use of both of these aspects of invariance.

The normative appeal of invariance should be obvious: to the extent that a relationship is more invariant, we can export or generalize or apply it to a range of different situations.[26] If causal cognition is well adapted to the achievement of goals having to do with generalizability, we would expect, as a descriptive matter, that it reflects the influence of invariance-related considerations. This provides a motive for looking empirically at whether human causal cognition reflects such influences.[27]

Cheng's causal power model

Cheng's causal power model (Cheng, 1997) makes use of a number of invariance assumptions and is intended both as a descriptive account of how subjects make causal judgments and of which judgments they make, but it also has a normative motivation—it is also intended as an account of how people ought to reason, thus illustrating our general theme of the interrelation between the normative and descriptive in understanding causal reasoning. It attempts to capture the intuitive idea that causes have "causal powers" that they "carry around with them" in different contexts. Cheng's model represents causes and effects as binary events, which can either be present or absent. Causes can be either "generative"—they can promote their effects—or they can be "preventive," interfering with the operation of generative causes. The "power" of a generative cause i to cause effect e is represented by p_i, the probability with which i causes e if i is present. Note that this is *not* the same as $P(e/i)$—among other considerations, the latter quantity reflects the influence of other causes of e that are present when i is.[28] Let a represent all such other causes of e, and assume that when present they produce e with probability p_a and that these are all generative rather than preventive causes of e. Assume also that e does not occur when it is not caused by either i or a. In a typical experiment, subjects have access to data about the frequencies of occurrence of i and e (e.g., in the form of a contingency table) but do not directly observe either the occurrence of a or p_i and p_a—these have to be inferred, to the extent that they can be. Cheng makes the following two additional assumptions about i and a:

8.1. i and a influence the occurrence of e independently.
8.2. The causal powers with which i and a influence the occurrence of e are independent of the probability with which i and a occur so that, for example, the probability that i occurs and causes e is just $P(i) \cdot p_i$.

Both (8.1) and (8.2) are invariance assumptions. (8.1) says that p_i is invariant under changes in p_a and similarly p_a is invariant under changes in p_i. (8.2) says that p_i and p_a are invariant across changes in the probability with which i and a occur. Cheng thinks of (8.1) and (8.2) as "default" assumptions that people bring to situations involving causal learning and judgment. I will return to the status of these, but the basic idea is that although nothing guarantees that such assumptions will be true in the situation of interest, they are nonetheless useful points of departure for reasoning which can be relaxed as the empirical evidence warrants.[29]

Given these assumptions, causal power can be represented and (in the appropriate circumstances) estimated in the following way:[30] First, $P(e)$ is given by the union of the probability that i occurs and causes e and that a occurs and causes e:

$$P(e) = P(i) \cdot p_i + P(a) \cdot p_a - P(i) \cdot p_i P(a) \cdot p_a$$

Conditionalizing on the presence of i, we obtain

$$P(e/i) = p_i + P(a/i) \cdot p_a - p_i P(a/i) \cdot p_a$$

Conditionalizing on the absence of i, we obtain

$$P(e/\text{not } i) = P(a/\text{not } i) \cdot p_a$$

Defining (8.3) $\Delta p(i) = P(e/i) - P(e/\text{not } i)$, it follows that

$$\Delta p(i) = p_i + P(a/i) \cdot p_a - p_i P(a/i) \cdot p_a - P(a/\text{not } i) \cdot p_a$$
$$= [1 - P(a/i) \cdot p_a] \cdot p_i + [(P(a/i) - P(a/\text{not } i)] \cdot p_a$$

Thus

$$p_i = \frac{\Delta p(i) - [P(a/i) - P(a/\text{not } i)] p_a}{1 - P(a) p_a}$$

Now consider the special case in which the probabilities with which a and i occur are independent so that $(P(a/i) = P(a/\text{not } i)$. Then

$$p_i = \Delta P(i) / 1 - P(a) p_a.$$

p_a cannot be estimated from the frequency data available, but since e is caused either by i or a, we can replace $P(a)$. p_a with $P(e/\text{not } i)$ yielding

$$p_i = \Delta p(i) / [1 - P(e/\text{not } i)] \qquad (8.4)$$

Thus under the specified assumptions, (8.4) is a normatively correct estimate for causal power, p_i.[31]

In many empirical studies (including Cheng's), subjects are presented with frequency information in some format about the patterns of co-occurrence between a candidate cause c and an effect e and are then asked to estimate (what is called) the "causal strength" of the relationship between c and e. Cheng's model claims that such causal strength judgments track p_i, causal power. As we shall see, there is disagreement about the verbal probe that is most appropriate for eliciting such judgments, but a commonly used question is some variant of "On a seven-point scale, how appropriate would it be to describe the relationship between c and e as one in which c causes e?" Cheng's model, as well as a number of other competing models, aim at (among other explananda) describing patterns of causal strength judgments and the representations and computations that underlie these. In many cases, what researchers aim to fit is something like average judgments across subjects—a practice that I will comment on below.

One of the main alternatives to Cheng's model is the so-called Δp model, according to which subjects' causal strength judgments will track the quantity (8.3) $\Delta p(i) = P(e/i) - P(e/\text{not } i)$. (This model has roots in associative models of animal learning such as the Rescorla-Wagner model.) As is apparent from (8.4), on the assumption that subjects' causal strength judgments track p_i, the predictions of Cheng's model and the Δp model diverge. Although there is nontrivial disagreement about the empirical facts (as will be discussed below), there is significant evidence favoring some features of the power pc model over Δp and other competitors.

To illustrate these diverging predictions, first consider situations in which $P(e/\text{not } i) = 1$. In this situation $\Delta p = 0$, since $P(e/i) = 1$, assuming i is a generative cause, the presence of which promotes e. Thus subjects guided by Δp in their causal judgments should report that in this situation, i does not cause e, assuming that a causal strength of zero corresponds to the absence of causation. By contrast, the denominator of (8.4) is zero when $P(e/\text{not } i) = 1$, so that (8.4) is undefined in this circumstance. Thus, subjects guided by p_i in their strength judgments should report that they are unable to reach any conclusion about the causal strength of i in this situation. Note that the latter judgment rather than the judgment based on Δp is the normatively correct one: when $P(e/\text{not } i) = 1$, a "ceiling effect" is present—since e is always present, the power (if any) of i to cause e cannot reveal itself in any differential probability of occurrence of e in the presence versus the absence of i. As an empirical matter, when ordinary subjects are given this option, a substantial number (but by no means all) chose this "unable to reach conclusion" alternative.

A second prediction, if subjects are guided by causal power p_i in their strength judgments, is this: as the probability of the effect $p(e)$ increases, $p(e/\text{not } i)$ will increase (assuming p_a. $P(a)$ is not zero). Thus, for a constant Δp, p_i will increase—in other words, i will be judged a stronger cause of e (again given the above assumptions) the more frequent the occurrence of e, for a fixed Δp. As an empirical matter, many subject judgments do tend to exhibit this feature, but this has often been treated as an "irrational bias" of some kind—it is certainly normatively inappropriate if the correct normative theory for judgments of causal strength is given by Δp. By contrast, this feature is both predicted when p_i is manipulated for constant Δp and shown to be normatively reasonable by the causal power theory.[32]

It is worth reflecting briefly on the normative differences between the causal power model and Δp. Intuitively speaking, Δp is normatively deficient as a measure of the causal strength of i because it does not correct for confounding—both $P(e/i)$ and $P(e/\text{not } i)$ reflect not just the relationship between i and e, but also the extent to which other causes of e, captured by a, are operative, even if those other causes operate completely independently of i. Among other limitations, this measure will not generalize appropriately to new situations in which the distribution of other causes of e is different than in the original situation. The invariance assumptions built into the causal power model correct for this (in effect by normalizing Δp to correct for other causes of e besides i), assuming the applicability of the assumptions that go into its derivation. In fact, in other experiments, Cheng and coauthors have shown that, as an empirical matter, causal power does a much better job of predicting which causal judgments (and associated measures of strength) generalize to new situations with different distributions of new causes than alternative measures like Δp—again, a pattern of judgment that is normatively reasonable.

I have said that although Cheng and others have obtained results supporting the empirical predictions of the model, the overall empirical adequacy of the model remains controversial. I turn now to a description of some discordant empirical results and the response of Cheng and her collaborators to these.[33] As we shall see, this discussion has a number of interesting philosophical and methodological implications. These include problems that can arise when verbal reports are used as evidence, both in experimental and armchair contexts, as well as issues having to do with subject heterogeneity.

Lober and Shanks (2000) agree that the causal power theory is the correct normative theory of causal judgment in the situations that satisfy the background assumptions of Cheng's theory. They draw attention, however, to two patterns present in human causal strength judgments that are prima facie inconsistent with the causal power model. The first consists in the fact that some subjects provide positive causal strength ratings in the presence of "noncontingency"—that is, when $\Delta p = 0$, with the magnitude of these ratings being influenced by P (e/not i). This is inconsistent with both the causal power and Δp models, which predict strength ratings of zero in such cases. Second, recall the experiments of Cheng that have been discussed, in which Δp is held constant, causal power is varied, and causal judgments are shown to track causal power. Lober and Shanks were able to replicate this result, but they also did the "opposite" experiment in which causal power is held constant across different experimental conditions and Δp varied. Of course, the causal power theory predicts no difference in judgment across these conditions but, averaging over the experimental population, such judgments *are* found to vary, appearing to support Δp over causal power. In an extremely interesting analysis of their data, Lober and Shanks show that their subjects can be separated into two groups, one of which (the power group) seems to be guided by the normative considerations that led to the construction of the causal power theory (e.g., this group is aware of ceiling effects and tries to take them into account in their causal strength judgments), and the other of which (the "contingency participants") seems not to take these considerations into account. When causal power is held constant across different values of Δp, the power participants behave pretty much as Cheng's theory predicts—their ratings are fairly constant across different values of Δp. By contrast, the ratings of the contingency group increase with increasing values of Δp, which is what the Δp but not the causal power model predicts.[34]

In an attempt to account for these and other results that appear to be inconsistent with the causal power model, Cheng and colleagues (Buehner, Cheng, and Clifford, 2003) appeal to several considerations (as well as further experiments). One has to do with what they call "ambiguity of the causal question." First, they note (following Tenenbaum and Griffiths, 2001) that standard verbal probes for causal strength (e.g., "how appropriate would it be to describe this as a situation in which c causes e?") may conflate a subject's degree of confidence that a causal relationship exists with the question of how "strong" that relationship is, given that it does exist. Second, they note a further potential ambiguity: when asked about the causal strength of c with respect to e, subjects might (i) interpret this as asking "what difference does the candidate cause make in the current learning context, in which alternative causes already produce e in a certain proportion of the entities?" (1126). Here the question is understood as asking, "what additional difference does c make to e, given that alternative causes are already causing some instances of e?" As should be obvious, Δp is the normatively correct answer to this question. A second, alternative way of interpreting the causal strength question is (ii) "what difference does the candidate cause c make in a context in which alternative causes never produce e?" where the normatively correct answer is given by the causal power model. Buehner et al. suggest that the results for experiments with positive contingencies that appear to be inconsistent with the causal power model might be explained by the fact that approximately half of the subjects are interpreting the causal strength question along the lines of (i) and the other half along the lines of (ii), and they provide an analysis of their data that supports this interpretation. They also performed a second experiment in which the causal question was altered along the lines of (ii), since (as they see it) this corresponds to the notion of causal strength that the power pc model is intended to capture. This revised causal question asked subjects to estimate "how many entities out of a group of 100 which did not show an outcome would now have the outcome in the counterfactual situation in which the candidate cause was introduced?" (p. 1128). As they note, this is in effect an "intervention" question, which asks what the effect of the cause would be if it were introduced by an intervention into a situation in which it was previously absent. Asking for an estimate in terms of proportion of entities also makes it clearer that the question is not about reliability or degree of confidence that a causal relation exists. Employing this revised verbal probe for causal strength and certain other modifications in their experimental design, Buehner, Cheng, and Clifford (2003) obtain results which seem to show that the great majority of their subjects judge in accord with the causal power model.

Some philosophical morals

Let me now try to extract some general philosophical morals for projects having to do with the empirical study of causal reasoning from this discussion and relate them to the ideas that have been previously discussed.

The central role played by normative ideas

Although Cheng's model is intended to account for empirical features of human causal judgment, it is motivated by normative considerations linking causal judgment to certain assumptions about the invariance properties of causal relationships. The link between the normative and the descriptive is provided by the claim that human causal judgment and learning are to some considerable degree normatively reasonable, so that normative models predict to some significant degree how people in fact learn and judge. This also illustrates the idea of "explaining success"—reasoning about causal relationships in terms of invariance (and adopting learning strategies that lead to invariant relationships, including estimating causal power from the relationship (8.4) rather than from Δp), contributes to success when this is understood as the discovery of relationships that are exportable to new situations.

Normative theorizing also enters in more subtle ways—for example, it motivates various experiments and the interpretation of experimental results. One would probably not think to do Cheng's experiments in which causal power is varied for constant Δp and the effect on judgment observed in the absence of a theory motivating the causal power model.[35] Normative theorizing also plays a role in interpreting the verbal probes used in eliciting strength judgments—for example, it is normative analysis that tells us that there is an important distinction between asking (i) what additional difference c makes to e in circumstances in which other causes of e are assumed to be present and asking (ii) what difference c would make to e if c were introduced in circumstances in which all other causes of e were absent. However, this role for normative theorizing in motivating the choice of verbal probe can introduce worries about a kind of circularity or lack of robustness, which I will discuss below.

I will add that this moral seems to me to generalize well beyond causal reasoning. In many cases, the most successful empirical or descriptive theories in the human sciences are those that are tied to normative theorizing—decision theory and theories of learning and belief change in response to evidence furnish additional examples. Normative theories can structure empirical investigation, motivate experiments, and help to interpret results. In philosophy of science, use of empirical data of any kind, whether it comes from experiments, surveys, case studies, or other sources, is likely to be most fruitful when connected to a normative theory. Moreover, the normative theory is not going to emerge just from the empirical data alone; instead, some independent rationale (typically connected to means/ends patterns of justification) is required.

The significance of intuitive judgments

In the experiments described in "Cheng's causal power model," people express judgments of causal strength when presented with various stimuli. Following a traditional armchair methodology, one might be tempted to interpret these as reports of intuitions that give us some sort of veridical insight into the nature of, or our concept of, causation, which the philosopher should then try to systematize. One obvious problem is that, as we have seen, different subjects have different and indeed inconsistent judgments about causal strength regarding the same cases. If these subject's judgments reflect intuitions, then at least in this case, the mere having of an intuitive judgment, no matter how firmly held, does not establish that such a judgment is correct or veridical. Nor is there any obvious reason to think that correctness of these intuitions can be established merely by systematizing them or bringing them into reflective equilibrium with one another. Instead, in the case under discussion, the normative correctness (or not) of intuitive judgments is established by an independently justified normative theory—the causal power model, which, in turn, is supported by various invariance assumptions.

A better way of thinking about the evidential significance of the judgments obtained in Cheng's experiments is simply that they tell us (or may tell us) something about what certain groups judge—this is certainly how Cheng understands their significance. We can then ask which models, normative or nonnormative, best account for such judgments. As noted, finding certain patterns in people's judgments may also prompt us to ask whether there may be some previously unconsidered normative rationale for those judgments, but it does not by itself show that there is such a rationale. As suggested above, I think that philosophers' appeals to intuitions should be treated in a similar way—such intuitive judgments can sometimes provide information about shared practices of judgment by others, just as Cheng's experimental results do, and we can then go on to ask whether there is some independent normative rationale for these judgments. When understood in this way, it is hard to see what grounds there are for a wholesale dismissal of appeals to intuition that would not in also be grounds for dismissal of the use of verbal judgments as a source of information in psychological experiments.[36] But it also follows that the mere having of an intuition is not normatively probative.

What people say versus why they say it

The research described above illustrates the important difference between reports of what people judge and the factors and processes that cause these judgments. Even if subjects accurately report their causal strength judgments, it seems clear that the representations, computations, and learning strategies that underlie those judgments (whether these are explained by the causal power model, some more associationist model incorporating Δp, or something else) are not themselves accessible (or capable of being established) via intuition. These are instead predicted on the basis of normative analysis and mathematical modeling and then require investigations—either experimental or observational—that control for confounders, for their confirmation.

Causal cognition and causal concepts

I have noted the tendency of philosophers following traditional intuition-based methodologies to frame their conclusions as claims about concepts (or their application), so that in the causation literature the primary conclusions such methods are taken to establish is that various scenarios do or do not fall under "our" concept of causation. As we see from the research described, there are many important features of causal cognition (or if you like, *causation*) that are not well captured in this way. To begin with, even if one holds that whether a causal relationship is present or not in some situation always requires a binary, yes-no judgment, it is clear that people make further more graded discriminations, distinguishing among causal claims with respect to how strong they are. Moreover, although I lack space for discussion, it is clear from other experimental and analytical work that "strength" has several distinct dimensions, some of which may be captured by the causal power model and others of which are not.[37] A complete theory of causation, whether normative or descriptive, should reflect this. In addition, there are a number of other features of causal reasoning that are not naturally viewed as constitutive of causal concepts but which are nonetheless important. These include the role played by various defaults, as I have illustrated. As we saw, Cheng's model incorporates various invariance assumptions such as the assumption that the tendency of i to cause e operates independently of the tendency of alternative causes a to cause e. Obviously such assumptions can be violated—causes can interact with one another to produce effects. (Indeed, Cheng devotes several papers to modeling how people reason about such interactive causes.)

Cheng's claim is that people tend to treat such invariance assumptions as defaults—in the absence of evidence to the contrary, they tend to first assume that causes operate noninteractively, only modifying their judgments and reasoning when they get evidence contradicting the original default. Such assumptions play the role of structuring inquiry (or search) in certain ways, leading inquirers to consider certain possibilities before others so that search among alternative hypotheses can proceed in an organized, systematic way. Because default assumptions can be violated and causal relations still can be present, it does not seem right to think of them as built into our concept of causation, but they still are important in causal reasoning. A similar remark applies to many other features of causal cognition. Ranging further afield, my guess is that philosophers of science interested in doing empirical work on "confirmation," "evidence," and related notions would also do well to focus less on what belongs to these concepts and more on the role of strategies of search, default assumptions, and the like. Again, there is no reason why empirical work relevant to scientific methodology should be organized around studies of concepts.

Potential ambiguity of the verbal probe

Issues surrounding the interpretation of verbal probes have important implications not just for the interpretation of experimental results but also for the role assigned to intuitive judgment in philosophical discussion. In an experimental context in which a question is posed (e.g., about causal strength or whether a causal relationship is present) and subjects are asked for a verbal response, it is obvious that one needs to worry that different subjects may interpret the question differently from one another or differently from what the experimenter intends. If so, the verbal probe may not measure what the experimenter thinks it is measuring or may not measure the same variable for all subjects.

It is natural to wonder whether the same thing sometimes may happen when philosophers elicit intuitions, either from themselves or others. Suppose Philo describes a case and reports having such and such an intuition about it. Cleanthes reports the same or a different intuition about the same case. Each is in effect asking themselves a question: ("What is my intuition or judgment about whether this is a case of X?") When (or how) can we be confident that they are asking themselves the same question? If Philo and Cleanthes both report their intuitions about the strength of the causal relationship present in a certain scenario (or whether a causal relationship is present at all), don't we need to consider the possibility that they may be interpreting causal strength (or "causal relationship") differently (as the subjects in experiments described above may be)? Indeed, going further, shouldn't we also be concerned about the possibility that Philo himself may be unaware of possible ambiguities in the question he is asking himself? He may report his intuition about causal strength (or whatever) without recognizing that the question he asks himself may be unclear or that when he asks himself what he thinks are versions of the same question, expressed slightly differently, he is actually asking himself different questions to which different answers are appropriate.[38]

As noted above, normative analysis of verbal probes can help to make us aware of possible ambiguities and unclarities in verbal probes, including those used in eliciting intuitions.[39] In addition, as illustrated above, additional empirical work can either confirm or disconfirm the possibility that different subjects are interpreting the same probe in different ways or differently from what the researcher intends. Still potential problems remain. When a normative model of causal judgment is invoked to support the use of a particular verbal probe, and the results of that probe are then used to support the descriptive adequacy of the normative model, there is an obvious worry about question-begging. For example, the revised verbal probe employed by Buehner et al. above appears attuned to direct subjects to formulate strength judgments just on the basis of the features that the causal power model claims do drive strength judgments. This does not make the fit between the model and the elicited judgments automatic or uninteresting (in fact, critics have claimed that the judgments elicited by this probe still do not fully track the predictions of the causal power model), but it does raise questions about how to assess the descriptive adequacy of a model which appears to be sensitive to the exact wording of the probe employed.

Should we be bothered by the possibility that if we were to employ a different verbal probe we would arrive at a different assessment of the adequacy of the model? At the very least, it seems that we should try to understand the relationship between different possible verbal probes and when, as an empirical matter, they elicit the same or different results and why this is the case. Going further, it would be highly desirable to combine verbal measures with measures of nonverbal behavior which can be less susceptible to concerns about misinterpretation. For example, in a causal learning task, the dependent variable might be whether the subject succeeds in activating a certain machine on the basis of presented information, with its being completely unambiguous whether this has been accomplished.[40] Many of the best designed and most persuasive experiments in causal cognition make use of nonverbal measures in part for this reason. I suggest that the same lessons can be applied in philosophy—to the extent that we employ reports of intuitions about cases (or surveys), we should consider whether different ways of eliciting the intuition generate the same or different results, whether the judgments generated are consistent with what is suggested by nonverbal measures, and so on.[41]

Implications for X-phi

I believe that much of what I have said about the role of intuitions (construed in the deflationary manner described above) transfers to survey-like X-phi investigations as well. Such surveys can be valuable in virtue of producing direct evidence bearing on how widely intuitions/judgments are shared, thus contributing to both negative, debunking programs and to more positive programs directed at the description of shared folk thinking. On the other hand, a number of the limitations of an intuition-based methodology are also potential problems for survey-style X-phi. We still have the problem that the verbal probes used in such surveys may be interpreted differently by different subjects or may contain unnoticed ambiguities. This is not, in principle, an insurmountable problem (just as it is not for reports of intuitions), and it is receiving more attention recently from experimental philosophers.[42] Nonetheless, it seems uncontroversial that experimental philosophers doing survey-style work should employ different verbal probes and try to understand the relationships between the results they produce. They should also try to understand what verbal responses tell us about nonverbal behavior and practices. In addition, as is the case with appeals to intuition, surveys by themselves will not tell us about the underlying causes of survey responses.

Finally there is another feature of survey-style research (and for that matter, traditional appeals to intuition) that deserves a brief mention. There is some tendency in this research to focus on the question of whether responses are very widely or nearly universally shared. On the one hand, this is a very natural and appropriate question when a philosopher claims, in an unqualified way, that "people judge that so and so . . . " On the other hand, to the extent that our interest is in normative theory and explaining success, discoveries that judgments and other practices are nonuniversal may be less consequential than is sometimes supposed. The fact (if it is a fact) that not all subjects judge in accord with the causal power model does not in itself undermine the normative status of that model or the normative appeal of invariance-based ideas. It is also consistent with the model providing an adequate explanation of the judgments of those subjects who do conform to the model and explaining why their causal cognition succeeds in certain respects. (Of course, some different explanation will be needed for those subjects who do not conform to the model.) To the extent that X-phi results are employed in a negative or debunking role (showing that many folk don't judge as philosophers claim), such results may not matter so much for projects of the sort outlined here.

Notes

1 Thanks to Richard Samuels and Daniel Wilkenfeld for helpful comments on an earlier version.
2 Of course I don't intend this as a criticism of experimental work on actual causation and related subjects.
3 It is of course possible to trivialize this sort of approach: for example, one might argue that the goal of causal thinking is to correctly describe what the causal facts are, thus rendering the approach completely unhelpful. What we want is a characterization of the goals associated with causal reasoning that gives us some independent purchase on when causal reasoning is correct or incorrect in virtue of contributing or not to these goals. "Describing the causal facts" is not a goal that can play this sort of role.
4 Of course one might adopt a more expansive conception of what is involved in our concept of (or the nature of) causation according to which this incorporates normative considerations, includes features of causal reasoning that are not analytically or constitutively part of the concept, and allows for the possibility that our concept may need clarification and re-engineering in various ways. Such conceptions will not be distinct from the normative project I describe.
5 See the discussion of "controlled" versus "natural" direct effects in Pearl, 2001.
6 Here I follow Knobe, 2016, who argues that a substantial amount of recent research in X-phi as well as in cognitive science is not organized around investigations into concepts and that, moreover, there are good reasons why it should not be.
7 I write "in a negative way" to accommodate the possible argument that intuitive judgments about cases as well as cognitive psychology show that the ordinary concept of causation is deeply confused, thus preparing the way for some very revisionary alternative account appealing to physics. Since I don't think there is any evidence for such confusion, I will ignore this possibility in what follows.

8   I will add that there are ways of construing the "what causation is" project that do not seem to me objectionable. For example, one might construe Cheng's causal power model, discussed below, as in some sense an account of what causation or at least what causal power is—an account according to which causal power is to be understood in terms of certain invariance assumptions. (One might also argue that such assumptions are part of the way we think about causation.) But this account is not the sort of thing metaphysicians are looking for when they ask what causation is—there is no accompanying story about special metaphysical entities or relationships that serve as truth makers for causal claims, no "reduction" of causation to other sorts of claims or anything along similar lines.

9   In other words, the idea is not just that causal reasoning allows us to successfully get around in the world but also that it enables a particular kind of success which is associated with manipulation and control.

10 Although the commonly accepted distinction between system 1 and system 2 reasoning is problematic in many respects, in light of these observations it is particularly problematic to assume system 2 reasoning is always normatively superior.

11 This too is a theme in much of the psychological research on causal cognition. See, for example, Gopnik, 2009.

12 It should be obvious that this sort of ends/means justification also depends on such considerations as whether the goals are coherent and achievable, whether the means proposed are such that they can actually be carried out by human beings, and so on. I take all of this to be built into the idea that such justification requires showing that the means *successfully* conduce to the goals.

13 Normative analysis is thus one point at which *a priori* or conceptual considerations legitimately enter into our story. Philosophers are not wrong to think that *a priori* reasoning has a role to play in thinking about causation, but they tend to mislocate that role, thinking that it has to do with the role of intuitions in delivering truths about the concept of causation.

14 This strategy is extremely common in the literature on causal cognition. In addition to the examples discussed below, see Gopnik et al., 2004 and Sobel et al., 2004.

15 For example, the important normative distinction between intervening and conditioning (Pearl, 2000) prompted experimental work showing that human causal cognition respects this feature. Notice, though, that both this and some of the possibilities discussed above are *not* a matter of empirical results providing "evidence" for the correctness of the normative theory. The interaction between normative and descriptive can take many other forms besides this evidential connection.

16 This process may also involve deciding that certain judgments are correct on the basis of systematic considerations even when people have no clear intuitions, as in Lewis' "spoils to the victor" arguments.

17 As an additional illustration of this difference, consider the contrast between attempting to justify some feature of our inductive practice in terms of reflective equilibrium versus justifying it in terms of means/ ends considerations. The reflective equilibrium strategy proceeds by collecting various inductive judgments we are inclined to make and then systematizing them, perhaps with further constraints. By contrast, in means/ends justification, one proceeds by specifying certain goals—for example, in the context of statistical inference, minimizing the probability of accepting false hypotheses and rejecting true ones—and then shows that certain testing procedures will achieve these goals. Intuitive judgments that cannot be justified in this way are rejected as mistaken, however deeply entrenched they may be and however much they cohere with other judgments. This, rather than appeals to reflective equilibrium, is the sort of justification procedure that is adopted in classical statistics.

18 The extent to which philosophers who report judgments about cases or intuitions are good judges of how others will judge is debated in the literature on intuition. In some cases, there is some evidence that philosophical and lay judgment diverge. In other cases, philosophers' judgments seem to accurately mirror lay judgment—see, for example, Nagel, 2016. Nagel's defense of appeals to intuition has important similarities with my own, since it rests on a "good judge of other's judgments" premise.

19 In other words, like Goldman, I don't take the content of the judgment to be what it is evidence for. In particular, I don't claim the overt content of the judgment necessarily has to do with shared practices.

20 The following example may help to clarify how this works. Consider the debates among philosophers of science in the 1950s and 1960s over the role of the directional or asymmetric features in explanation. Critics of the deductive nomological (DN) model, such as Scriven, argued that the model was defective because it did not capture these features, mistakenly allowing the length of the shadow cast by a flagpole to explain the length of the pole. In a recent discussion, Stich and Tobia, 2016 treat this an example of an appeal to intuition, with the DN model being rejected because it is contrary to our intuitions about the flagpole case. I agree with Stich and Tobia that it is not justifiable to reject the DN model merely because it is contrary to intuition in this way. On the other hand, the intuition we have about the example brings to our attention that our practices of explanatory judgment are such that directional considerations play an important role. This can motivate us to ask whether there is some normative basis for distinguishing between flagpole-to-shadow explanations and the reverse. Although I won't argue

in their support here, several recent accounts claim to provide such a normative basis. In appealing to such accounts, we don't conclude that the original intuition is correct merely because people have it—rather, its correctness derives from a normative theory that supports it.

21 I also make no claims about whether a lot of X-phi research that is survey-like suffers from methodological flaws qua surveys. Again, in distinguishing between bare surveys and causal analysis, I don't mean to claim that the surveys employed in X-phi and elsewhere are generally bad or flawed surveys—my point is rather that there are important limitations on what even a well-designed survey can accomplish.

22 One standard typology distinguishes between the negative and positive program in X-phi. The negative programs appeals to empirical results about how ordinary subjects judge to undermine armchair philosophical claims about how most people judge and claims about the nature of "our" concepts based on such claims. The positive program attempts to use claims about how most people to judge to provide evidence for how the folk think about such and such or for claims about the structure of folk concepts. Note that in both cases survey-like results seem to be all that is required to accomplish these ends. To the extent that some significant portion of X-phi research falls into either of these two categories, this helps to explain why it is (legitimately) survey-like.

On the other hand, although I have not tried to gather systematic evidence for this claim, it is a plausible conjecture that although a substantial amount of X-phi research was once organized around investigations, negative or positive, into claims about concepts, this is much less true of more recent X-phi research, some significant portion of which looks much like cognitive science (as Knobe, 2016, argues) and shares its goals, which include causal analysis. In any case, I make no claims about how much of X-phi falls into the categories I distinguish; what matters for my purposes is the distinctions themselves.

23 Thus the difference between surveys and causal analysis is not that the former is observational and the latter is experimental. What is distinctive of a survey is that the design and the observations made are not such that they support causal analysis.

24 That something else besides the fact that no connecting process is present influences causal judgment in disconnection cases is strongly suggested by the fact that, typically, subjects distinguish *among* dependence claims in which there is no connecting process, assigning some greater causal strength than others.

25 In Woodward, 2006, following David Lewis, I used "insensitive" to describe this particular aspect of invariance—that is, invariance under changes in background conditions.

26 I have discussed invariance related notions in a number of different places (2003, 2006, 2010, 2018) and draw on this discussion for motivation in what follows.

27 In addition to the research by Cheng described here, other recent empirical research that highlights the role of invariance in causal reasoning includes Lombrozo, 2010; and Vasilveya, forthcoming. Kominsky et al. (2015) and Icard (forthcoming) highlight a role for invariance in actual cause judgment.

28 Thus within this framework, "cause" is not defined in terms of probabilities, although it is assumed that there are systematic relationships between these notions.

29 I see this notion of defaults as a useful way of thinking about the status of methodological maxims in science more generally.

30 Of course Cheng does not assume that subjects consciously reason in accord with the algebra that follows— what she presents is a computational level, rational reconstruction, with subjects judging as if they computed and represented causal power in the manner described below. Subjects have access to their judgments but typically not to the processes that produce those judgments.

31 Note that causal power is not *defined* as (8.4). (8.4) provides a formula for computing causal power when certain specific additional assumptions are satisfied. When these assumptions are not satisfied, causal power may still be well defined but it is not identifiable from the data.

32 The fact that judgments of causal strength of i with respect to e increase as $P(e)$ increases is a striking example of an observation that will initially seem normatively unreasonable to many but which has a nonobvious normative rationale. This is thus an illustration of the claim in the section "Relating the normative and descriptive" that observing that certain judgments occur, as a descriptive matter, may prompt a search for a normative theory that makes sense of them.

33 What follows is, for reasons of space, a very partial and incomplete description of a complicated empirical situation.

34 To the extent that such subject heterogeneity is real, merely reporting average subject behavior across an entire experimental population omits important information and can be quite misleading. Of course this is a general methodological problem in psychology.

35 A similar point holds for the experiments described in Vasilyeva et al. (forthcoming) that test whether subject causal judgments are influenced by other sorts of invariance features of a sort described in Woodward, 2010.

36 To (so to speak) turn this point around, if (as I assume is likely) you are not tempted to think that the causal strength judgments made by the subjects in Cheng's experiments are evidence for the truth of those judgments or sources of rational insight into the nature of causation, you should adopt a similar stance towards the intuitive judgments of philosophers.

37 See Woodward, forthcoming, for additional discussion.

38 For example, in considering whether C causes E, one needs to distinguish the question of whether C has a nonzero net effect on E, from whether C causally contributes to E along some route. When there is cancellation along different

routes, these questions should receive different answers. A look at the philosophical discussion around, for example, Hesslow's birth-control pills example (Hesslow, 1976) will show that many philosopher's judgments about this example failed to note this distinction.

39 It is worth noting explicitly that this is the kind of task (noting ambiguities, making distinctions) which traditional philosophers are good at. Properly conducted empirical analysis will thus be dependent on this sort of work as well as on normative analysis more generally.

40 As in Gopnik et al., 2004.

41 I don't have worked-out ideas about how to do this, but there are suggestive examples from the psychology literature: researchers have looked at the relationship between verbally expressed judgments of causal strength or power and measures having to do with subject's willingness to select one cause rather than another to bring about some desired goal, willingness to generalize to new situations as evidenced in nonverbal behavior, and so on.

42 Such ambiguity can often be detected by a combination of analysis and additional experimentation, as we see illustrated by the research by Cheng and Shanks described here.

References

Angrist, J. and Pischke, S. (2009). *Mostly Harmless Econometrics*. Princeton: Princeton University Press.
Buehner, M., Cheng, P., and Clifford, D. (2003). From covariation to causation: A test of the assumption of causal power. *Journal of Experimental Psychology: Learning, Memory, and Cognition*, 29: 1119–40.
Cheng, P. (1997). From covariation to causation: A causal power theory. *Psychological Review*, 104: 367–405.
Dowe, P. (2001). *Physical Causation*. Cambridge: Cambridge University Press.
Goldman, A. (2007). Philosophical intuitions: Their target, their source, and their epistemic status. *Grazer Philosophische Studien*, 74: 1–26.
Gopnik, A. (2009). *The Philosophical Baby*. New York: Farrar, Straus and Giroux.
Gopnik, A., Glymour, C, Sobel, D., Schulz, L., Kusnir, T., and Danks, D. (2004). A theory of causal learning in children. *Psychological Review*, 111: 1–30.
Hesslow, G. (1976). Two notes on the probabilistic approach to causality. *Philosophy of Science*, 43: 290–92.
Hitchcock, C. and Knobe, J. (2009). Cause and norm. *Journal of Philosophy*, 106: 587–612.
Holyoak, K. and Cheng, P. (2011). Causal learning and inference as a rational process: The new synthesis. *Annual Review of Psychology*, 62: 135–63.

Icard, T., Kominsky, J., and Knobe, J. (2017). Normality and actual causal strength. *Cognition* 161: 80–93.

Lewis, D. (1986). *Philosophical Papers* (Vol. 2). Oxford: Oxford University Press.

Kamm, F. (1993). *Morality, Mortality* (Vol. 1). New York: Oxford University Press.

Knobe, J. (2016). Experimental philosophy is cognitive science. In J. Sytsma and W. Buckwalter (Eds.), *A Companion to Experimental Philosophy* (pp. 37–52). Malden, MA: Wiley Blackwell.

Kominsky, J., Phillips, J., Gerstenberg, T., Lagnado, D., and Knobe, J. (2015). Causal superseding. *Cognition*, 137: 196–209.

Lober, K. and Shanks, D. (2000). Is causal induction based on causal power? Critique of Cheng (1997). *Psychological Review*, 107: 195–212.

Lombrozo, T. (2010). Causal-explanatory pluralism: How intentions, functions, and mechanisms influence causal ascriptions. *Cognitive Psychology*, 61: 303–32.

Machery, E., Mallon, R., Nichols, S., and Stich, S. (2004). Semantics: Cross-cultural style. *Cognition* 92: B1–B12.

Mortensen, K. and Nagel, J. (2016). Armchair- Friendly experimental philosophy. In J. Sytsma and W. Buckwalter (Eds.), *A Companion to Experimental Philosophy* (pp. 53–70). Malden, MA: Wiley Blackwell.

Paul, L. and Hall, N. (2013). *Causation: A User's Guide*. Oxford: Oxford University Press.

Pearl, J. (2000). *Causality*. Cambridge: Cambridge University Press.

Schaffer, J. (2000). Causation by disconnection. *Philosophy of Science*, 67: 285–300.

Sobel, D., Tenenbaum, J., and Gopnik, A. (2004). Children's causal inferences from indirect evidence: Backwards blocking and Bayesian reasoning in preschoolers. *Cognitive Science*, 28: 303–33.

Stich, S. and Tobia, K. (2016). Experimental philosophy and the philosophical tradition. In J. Sytsma and W. Buckwalter (Eds.), *A Companion to Experimental Philosophy* (pp. 5–21). Malden, MA: Wiley Blackwell.

Tenenbaum, J. and Griffiths, T. (2001). Structure learning in human causal induction. In T. Leen, T. Dietterich, and V. Tresp (Eds.), *Advances in Neural Processing Systems* (Vol. 13, pp. 59–65). Cambridge: MIT Press.

Tenenbaum, J., Griffiths, T., and Kemp, C. (2006). Theory-based models of inductive learning and reasoning. *Trends in Cognitive Science*, 10: 309–18.

Turri, J. (2016). Knowledge judgments in "gettier" cases. In J. Sytsma and W. Buckwalter (Eds.), *A Companion to Experimental Philosophy* (pp. 337–48). Malden, MA: Wiley Blackwell.

Vasilyeva, N., Blanchard, T., and Lombrozo, T. (2018) Stable causal relations are better causal relations. *Cognitive Science* 42: 1265–1296.

Walsh, C. and Sloman, S. (2011). The meaning of cause and prevent: The role of causal mechanism. *Mind and Language*, 26: 21–52.

Woodward, J. (2003). *Making Things Happen*. New York: Oxford University Press.

Woodward, J. (2006) Sensitive and insensitive causation. *The Philosophical Review* 115: 1–50.

Woodward, J. (2007). Interventionist theories of causation in psychological perspective. In A. Gopnik and L. Schulz (Eds.), *Causal Learning: Psychology, Philosophy and Computation* (pp. 19–36). New York: Oxford University Press.

Woodward, J. (2010). Causation in biology: Stability, specificity and the choice of levels of explanation. *Biology and Philosophy* 25: 287–318.

Woodward, J. (2018). Laws: An invariance- based account. In W. Ott and L. Patton (Eds.), *Laws of Nature* (pp. 158–80). Oxford: Oxford University Press.

Woodward, J. (Forthcoming). *Causation With a Human Face*.

Contributors

Mark Fedyk is an associate professor in the Department of Philosophy at Mount Allison University and a research associate at The Ottawa Hospital Research Institute. His research covers topics in ethics, the philosophy of science, and the cognitive sciences. He recently published *The Social Turn in Moral Psychology* (2017).

Kyle Fricke is a curriculum development analyst at the University of Texas at Austin and has received an MA in science and mathematics education from the University of California, Berkeley. He has published other work (e.g., in 2017) relating to climate change and education.

Paul Griffiths is Australian Research Council Laureate Fellow and professor of Philosophy at the University of Sydney, Australia. His recent publications include (with Karola Stotz) *Genetics and Philosophy: An Introduction* (CUP 2013)

Frank Keil is the Charles C. and Dorathea S. Dilley Professor of Psychology, Linguistics and Cognitive Science at Yale.

Elizabeth Kon is a cognitive psychologist operating out of UC Berkeley, and she has had papers presented and published in the proceedings of the Cognitive Science Society Conference.

Paras Kumar is a graduate student in science and mathematics education at the University of California, Berkeley. His undergraduate thesis for his Environmental Science BS (also at the University of California, Berkeley) assessed how communicating science data using graphs reduced, without polarization, climate skepticism. He is also the founder and owner of *debatedrills.com*, a debate coaching organization.

Tamar Kushnir is an associate professor in the Department of Human Development at Cornell University.

Lee Nevo Lamprey received an MA in science and mathematics education from the University of California, Berkeley. She co-created

HowGlobalWarmingWorks.org and, among other works, she co-authored *Increased Wisdom from the Ashes of Ignorance and Surprise: Numerically-Driven Inferencing, Global Warming, and Other Exemplar Realms* (Ranney, Munnich, & Lamprey, 2016).

Stefan Linquist is an associate professor in the Department of Philosophy and adjunct professor in the Department of Biology at the University of Guelph. His main areas of research include the philosophy of ecology, cultural evolution, conceptions of human nature, and the philosophy of genomics. He publishes in such journals as *Philosophy of Science, Biology and Philosophy*, and *The British Journal for the Philosophy of Science*, and he is coauthor (with Gary Varner and Jonathan Newman) of *Defending Biodiversity: Environmental Science and Ethics* (CUP 2017).

Tania Lombrozo is a professor at Princeton University in the Department of Psychology. Her work aims to address foundational questions about cognition using the empirical tools of cognitive psychology and the conceptual tools of analytic philosophy. Her work focuses on explanation and understanding, conceptual representation, categorization, social cognition, causal reasoning, and folk epistemology.

Edouard Machery is Distinguished Professor in the Department of History and Philosophy of Science at the University of Pittsburgh and the director of the Center for Philosophy of Science at the University of Pittsburgh. His research focuses on the philosophical issues raised by psychology and cognitive neuroscience, and he has also extensively contributed to experimental philosophy. He is the author of *Doing without Concepts* (OUP, 2009) and *Philosophy Within Its Proper Bounds* (OUP, 2017).

Michiru Nagatsu is Associate Professor in Practical Philosophy and Helsinki Institute of Sustainability Science at the University of Helsinki, Finland. He works on the foundations of human sociality and on the methodology of interdisciplinary research, in particular sustainability and behavioral sciences. His most recent book (edited with Attilia Ruzzene) is *Contemporary Philosophy and Social Science: An Interdisciplinary Dialogue* (2019).

Michael Andrew Ranney is a professor at the University of California, Berkeley (serving Education, Psychology, and Cognitive Science, etc., faculties). He has headed Berkeley's Reasoning Research Group since a postdoctoral fellowship at Princeton's Cognitive Science Laboratory, which followed experimental cognitive psychology PhD/MS degrees succeeding undergraduate microbiology and psychology majors. Ranney studies/fosters public (relative) coherence—lately about climate change—and designs thinking-improving curricula/methods/software (e.g., HowGlobalWarmingWorks.org). Ranney was a National Academy of Education (and Spencer Foundation) Spencer Fellow and a UC Regents' Junior Faculty Fellow. A Psychonomic Society Fellow, he has published extensively in diverse fields (including Applied Physics, Animal Learning, Communication, etc.).

Richard Samuels is a professor of Philosophy and Cognitive Science at the Ohio State University, and he has previously held appointments at King's College London and the University of Pennsylvania. He has published extensively on topics in the philosophy of psychology and the foundations of cognitive science.

Matthew Shonman is an analyst at the Cybersecurity and Infrastructure Security Agency. Matthew holds a BA in cognitive science from the University of California, Berkeley, and an MS in security informatics from Johns Hopkins University. As a researcher, he has explored both the effectiveness of science communication methods and the cognitive processes that underlie phishing email detection. Matthew's most recent publication is *Simulating Phishing Email Processing with Instance-Based Learning and Cognitive Chunk Activation* (2018).

Andrew Shtulman is a professor of Psychology and Cognitive Science at Occidental College. He studies conceptual development and conceptual change, particularly as they relate to science education. He is the recipient of an Early Career Development Award from the National Science Foundation and an Understanding Human Cognition Scholar Award from the James S. McDonnell Foundation, and he is the author of *Scienceblind: Why Our Intuitive Theories About the World Are So Often Wrong* (2017).

Karola Stotz is senior lecturer in Philosophy at Macquarie University, Sydney, Australia. Her recent publications include (with Paul Griffiths) *Genetics and Philosophy: An Introduction* (CUP 2013)

Daniel Wilkenfeld is a philosopher of science operating out of the University of Pittsburgh. He works in a variety of areas, including philosophy of science, experimental philosophy, and bioethics/philosophy of disability. He has publications in journals such as *Philosophical Studies* and *Synthese*.

James Woodward is Distinguished Professor in the Department of the History and Philosophy of Science at the University of Pittsburgh and J. O. and Juliette Koepfli Professor Emeritus at the California Institute of Philosophy. He works mainly in general philosophy of science, and much of his recent work has been on causal reasoning. His 2003 book *Making Things Happen: A Theory of Causal Explanation* won the 2005 Lakatos Award. He is a fellow of the American Academy of Arts and Sciences.

Fei Xu is a professor of psychology at UC Berkeley. She is internationally known for her work on cognitive and language development. She is a fellow of the Association for Psychological Science and the Cognitive Science Society, and a 2018 Guggenheim Fellow.

Index

abstract concepts   125–6, 127–8
accuracy monitoring, child's theory of evidence   129–30
acquired traits   172
adaptation of species   110–11
advancement of science, and knowledge   98
ambiguities   210, 229–30
American Association for the Advancement of Science (AAAS)   97
analogous traits   108, 113
Angner, E.   157–8, 159–60, 164, 167 n.7
animal natures *see* birdsong study; three-factor model of animal natures
anomalies   103–4, 157
Anthropocene extinctions   89
apparent mechanistic ignorance   44–5
a priori reasoning   238 n.13
Aristotle   172
artificial learning tasks   31, 32
astronomers vs. nonastronomers   104
astronomy, anomaly of   104
attractor model   181, 191–2, 196
Auspurg, K.   166 n.4

background assumptions   104–6
Baker, M. C.   174
bare surveys   217–19, 240 n.21
Bechtel, W.   43
behavioral change   152–3, 154, 155, 162, 163–4, 166 n.4, 167 n.5
behavioral economics   148, 155, 157, 161, 162, 163
behavioral trait   111
behaviorism   156, 159, 160
behaviorist myth   156, 157
belief change   152, 166 n.4, 231
beliefs, and theories   100
better rule reporting   21–2, 23–6, 29–30
biological traits   173, 176
biologists vs. nonbiologists   106–7, 108

biology   110
  errors of omission   99
  mechanistic explanations in   42–3
  philosophy of   5, 11
birdsong study   177–8, 197
  biologists' responses   186
  evolutionary behavioral scientists' responses   188
  experimental psychologists' responses   188
  generative linguists' responses   189
  linguists' responses   186
  mixed-design analysis of different responses   186–90
  nongenerative linguists' responses   189–90
  psychologists' responses   185
Black's theory of caloric   103–4
boundless growth, as a danger to ecological systems   90
Bovens, L.   164, 167 n.6
Boyd, R.   174
Buehner, M.   230, 235

Canada, nationalism and global warming acceptance in   91 n.7
Carey, Susan   173
Carpenter, K.   65
Cartesian skepticism   139–40 n.2
casual observation   107
causal abstractions   48–9
  and division of cognitive labor   50
  and relearning   49
  and self-understanding   48
  and sense of relevancy   48
  and sense of relevant experts   48
  and validation of expert's testimony   48–9
causal analysis   220–4
causal cognition and causal concepts   233–4
causal complexity   47, 48, 49

causal explanation 206
causal-explanatory theories 136–7
causal inquiry 46
causal learning 127, 131
causal learning, verbal probe 235–6
causal modeling, confounding factors 221–2
causal networks, and mechanistic understanding 51–2
causal power theory 210, 211, 225–30, 236–7
  alternatives to 227–8
  assumptions 225–6
  as normative theory of causal judgment 229
  representations and computations 226–7
causal reasoning 205
  causal cognition and causal concepts 233–4
  common-sense 210
  and coping ability 208–9
  descriptive theories of 206
  and goals 210–11
  intuitive judgments 232
  nontrivial heterogeneity in 207–8
  normative ideas 231–2
  people's judgment (reports, factors and processes) 233
  potential ambiguity of the verbal probe 234–6
causal relationships
  and counterfactuals 207
  invariance 211
  invariance assumptions 210, 221, 224–5, 240 n.25
  varieties of 207
causal strength 230
causal thinking
  goal of 237 n.3
causation 206, 238 n.8
  distinction between concept and underlying nature 208
  as normative, philosophical theories of 207
  "our" concept of 207–8, 209, 216, 232, 233, 234, 240 n.22
  philosophy of 11
cause, being aware and unaware of 152

causes and effects 225–33
Chang, H. 162
changing methodological practice 161–5
chemists vs. nonchemists 102, 103–4
Cheng, P. 206, 210, 211, 228–9, 230, 241 n.30 see also causal power theory
children
  ability to reason about context and goals interaction with judgments 126–7
  causal abstractions in 48, 49
  classification of information as evidence 126–7
  construction of knowledge 123–4
  early attention to source of information 133–4
  estimations of relevancy of information to learning 132–3, 139 n.2
  evidential concepts 123–4, 128, 134
  factual versus casual explanations 46
  intuitions about causal complexity of devices 47
  mechanistic understanding, interest in 42, 44, 45–7, 53
  origins of understanding in 42
  relative expertise assessment by 131–2
  theory of evidence (see theory of evidence)
choice 155, 165 see also preference
  concepts
  concepts 151–5
  consciousness 152
choice behavior, as utility maximization 156
choice theory 151–2, 154
  to individual humans again 159–60
  and preferences 156
Chomsky, Noam 174
Clark, D. 64, 65, 72, 73–4, 82, 83, 84, 87, 91 n.9
Clifford, D. 230
cognitive activities 1, 7, 98, 100
cognitive foundations, of science 8–9
cognitive psychology as epistemology 134–8
cognitive science, and mechanistic understanding 41, 42, 44
Cohen, J. 83

Cohen, S. 83
commissions, environmental
  perceptions 99–100
commonsensible realism 150–1, 152, 154, 155, 159, 164, 165
community of knowledge effect 50–1
concepts 136
  abstract concepts 125–6, 127–8
  epistemological concepts 138
  normative concepts 138
conceptual diversity 7–8
conceptual ecology 149, 162, 163, 166, 180, 181, 191–2
conceptual uniformity 8
conceptual variance 149, 152
confounding factors 220–1
consciousness 2, 152
consciousness, and choice 152
conservativeness of economics 160–1
constructive preferences 160–1
contemporary revealed preference theory 162
contexts, and judgment 126–8
continental drift, background assumptions 105
cooperative behaviors 112
Cope's theory of accelerated growth 108
counterevidence and counterinstruction, resistance to 114
counterintuitive
  and background assumptions 104–5
  science as 98
Cowen, T. 162
Craver, C. 43
creationism 107
cultural attractor theory 181
culture
  and beliefs 90–1 n.2
  versus cognition 65
  versus information 65
  information transmission 108
  and theory change 114
cumulative prospect theory 167 n.9

Darden, L. 43
Darwin's theory 109–11, 112
decision theory 231
deductive nomological (DN) model 239 n.20

default assumptions 233–4
descriptive theory 206, 209–10
developmental fixity, traits 173
disciplinary classificatory scheme 184
disconnection scenario 222
discovering vs. learning scientific truths 101–4
  divergent background assumptions 104–6
  divergent empirical concerns 103–4
  divergent explanatory goals 101–3, 106–7
  divergent paths to understanding evolution 106–13
  understanding evolution (*see* evolution, understanding of)
divine creation 107
domain-specific theories 123
Dowe, P. 219, 222
Dweck, Carol 73

economist effect 154, 167 n.5
economists 153
Edwards, B. J. 37 n.6
Edwards, J. 156
empirical adequacy of the model 228–9
empiricists
  attribution of people 64–6, 90 n.2
  use of traits by 174
ends/means justification *see* means/ends justification
epistemic inferences 46
epistemologists 2, 139–40 n.2
epistemologized psychology 124, 134–8
  future research in 138–9
epistemology
  cognitive psychology as 134–8
  intuitive epistemology (*see* theory of evidence)
  naturalization of 137, 138
errors of commission 99–100
errors of omission 99
essentialism 99
evolution, understanding of
  divergent background assumptions 109–13
  divergent empirical concerns 108–9
  divergent explanatory goals 106–7

evolutionary behaviors, and innateness 174, 188
evolutionary philosophy of science 180
evolutionary trees (misleading information) 108–9
exceptionless generalizations 27, 31
exceptionless laws 19, 21, 25, 27
exceptionless patterns 16, 17, 18, 19–20
expected utility theory 156
experimental economics of philosophy 148
experimental philosophy 3, 213, 223–4
 advantages of 165
 cognitive foundations 8–9
 and conceptual diversity 7–8
 and conceptual uniformity 8
 contributions of 7–9
 definition of 3–5
 of economics 148, 151, 159, 165–6
 ecumenical conception 4–5
 and empirical methods 4
 as an extension of STS 5–6
 as an extension of the "turn to practice" 6–7
 folk vs. economic preferences 159
 implications for 236–7
 negative and positive program in 240 n.22
 neutrality 5
 of psychology 159
 and scientists' concepts 8
 subject matter and methodology 3–4
 survey-like 236, 240 n.21
experimentation, counfounding factors 221
explaining for the best inference (EBI) 31
explanation 9, 15
 better rule reporting 21–22, 23–26, 29–30
 in control condition 16–17
 effects of engaging in 31–2
 frugal explainers 17
 and generalizations involving exceptions 20–7
 ideal explanations 31
 ideal rule and no ideal rule conditions 21, 22–26, 29–30, 38 n.9
 and learning environment 16, 27–31
 limitations of experiments of 31–2
 patterns-type 15–16, 17–20, 21, 22–6, 29–30, 38 n.9
 prompt-type 16–17, 18–20, 21, 22–3, 25–9, 30, 37–8 n.9
 role in learning 15–20
 rule ratings 22, 24, 26–27
 rule reporting 23, 28
 satisfying explanations 16–17
 worse rule reporting 21, 22–23, 24, 26, 27, 30
explicit reasoning 210
external anatomical trait 111
extinct species 108

facts 215
fake news 92 n.12
fixed mindset 73
fixity 173, 177–8, 180, 182, 183, 185, 186, 188–90, 193, 194–5, 197–8
Fodor, J. A. 174
folkbiology 172–3, 176, 179, 196
folk concept of preferences 159
 vs. economic choice 151–5
 vs. scientific variance 149
functional traits 173

game theory 157
Garcia de Osuna, J. 70
generalizations 19–20
 exceptionless 27, 31
 involving exceptions 20–7
generative linguists 190
generics 176
genetic essentialism 179, 180
geologic time, participants' understanding of 112
geologists vs. nongeologists 105
global climate change
 denial of 66–7
 graphs spawn scientific inferences 70–1
global warming acceptance
 changed beliefs 74–5
 change with representative and misleading statistics 72–6
 changing beliefs with supra-nationalistic and super-nationalistic statistics 76–82

durability using mechanistic
    explanations   72
feedback statistics   72–3
gains   70, 72, 84, 85, 86, 87
with graphs and averaging   67–72
mechanism, ignorance of   89
with no polarization   71, 75–6, 79, 82,
    86, 87, 88
posttest   68, 69–70, 73, 74, 75, 78,
    79–80, 82, 84–7
pretest   68, 70, 71, 73, 75, 78, 79–80,
    82, 84
with scientific statistics, texts, and
    videos   82–7
goals
    and causal judgment   210–11
    and judgment   126–8
Goldman, A.   215–16
Gopnik, A.   206
Griffiths, Paul   7, 166 n.3, 176, 180, 194
growth mindset   73
Guala, F.   158, 159, 160
Gul, F.   157

Hall, N.   214
Harreveld, F.   81
Hausman, Daniel   150, 157, 160, 161, 164
Havranek, M.   88
heat, anomaly of   103–4
Hempel, Carl   9, 15
Hesslow, G.   242 n.38
Hinz, T.   166 n.4
Hitchcock, C.   223
homologous traits   108
HowGlobalWarmingWorks.org   89, 90
Hull, D. L.   180
human concerns, of scientific
    research   15
human visual system   209

Icard, T.   223
idealization   150, 161, 167 n.11
ideal rule condition, pattern-type   21,
    22–6, 29–30, 38 n.9
illnesses, background assumptions   106–7
illusions
    of having more knowledge   50
    illusion of explanatory depth
        (IOED)   45, 50, 53
    of understanding   45

impetus theory   100
incentives   163–4, 166
inferences
    epistemic   46
    and theory of evidence   128
innate ideas   114
innateness   172–3 see also three-factor
        model of animal natures
    attractor model   181, 191–2, 196
    birdsong study (see birdsong study)
    criticism of   173
    and evolutionary behaviors   174
    in science   196–8
    scientists' effective concept of   175
    traits   172–3, 176, 197–8
    vernacular concept of   176–9, 196
intention   2, 125, 131
interdisciplinary variance   149
internal anatomical trait   111
internet search   49–50
interpersonal learning   127
interspecific behaviors   112
intervening and conditioning, distinction
        between   238
intradisciplinary variance   149
intransitive preferences   158, 159
intraspecific behaviors   112
intraspecific competition, understanding
        of   111–12
intuitions   2, 3, 7, 50, 213–14
    and acquired knowledge   215
    of children   46, 47, 50
    intuitive judgments   223, 232,
        239 n.17
    intuitive theories of evolution
        99–101, 100, 114, 115
    limitations   216–17
invariance, causal relationships   210, 221,
        224–5, 240 n.25
irrational bias   228

judgments   216 see also causal
        reasoning
    of causal strength   241 n.32
    intuitive judgments   223, 232,
        239 n.17
justification   2

Kahan, D. M.   10, 65, 82
Kahneman, D.   167 n.11

Knobe, J.   223, 237 n.6
knowledge   2, 86
    acquisition of   10–11
    outsourcing, consequences of   49–51
knowledge-score gains   86
Kominsky   223
Kon, Elizabeth   17, 18, 19–20
Kuhn, Thomas   10

labeling   126, 129, 131–2, 134
Lamarck's theory of acquired
    characters   100
Lamprey, L. N.   82
learning   139–40 n.2, 212
    causal learning   127, 131
    children's estimations of relevancy of
        information to   132–133, 139 n.2
    environment, effects of explanation
        in   16, 27–31
    interpersonal learning   127
    theory of   231
    and theory of evidence   123, 127,
        132–3, 139 n.2
Lee, H. S.   112
Lewandowsky, S.   67, 68
Lewis, David   139–40 n.2, 207, 214, 219,
    239 n.16
Linquist, S.   166 n.3, 194
Lober, K.   229
Lombrozo, Tania   16, 17, 18, 19–20, 31,
    37 n.6, 222
loyalty   91 n.4

Machamer, P. K.   43
Machery, Eduoard   8, 166 n.3, 194, 220
Mäki, Uskali   150
manipulation   211, 221
market structure   160
matter, discovering vs. learning
    about   101–2
means/ends justification   211–12, 238
    n.12, 239 n.17
mechanism metadata   47
mechanistic interventions   82–4, 86
mechanistic understanding/
        explanations   9–10, 42
    apparent absence of   44–5
    in biology   42–3
    and causal abstractions   48–50
    in children   45–7

emergence of   42–4
    in folk science   44, 53–4
    functional language   43
    future inquiry   51–2
    of global warming   72, 82, 87, 88
    hierarchical property of   43
    and higher-order abstract principles
        learning   47
    and knowledge outsourcing   49–51
    patterns of interactions and
        higher-level process   43
    subassemblies property   43
    transfer of energy   43
mental theory   126
Mill, John Stuart   62, 63, 77, 87, 88,
    89–90
    on new knowledge   66
    science of national character   63, 89
mindset
    changes in   75
    fixed vs. growth   73
minimalism   158
misleading intervention   72–6
mixed-effects models   75, 80
motion, discovering vs. learning
    about   102–3
Munnich, E. L   82

Nagatsu, M.   164
Nagel, Ernest   159
nationalism   63, 66, 88
    and global warming acceptance   76–82
    pro-nationalism   79, 81
    reduction of   79
    super-nationalism   76–82, 88
    supra-nationalism   76–82, 86, 88
    supra-nationalistic and super-
        nationalistic statistics   76–82
natural selection   109–11, 112, 212
negative program   240 n.22
Nelson, J.   70
neuroeconomics   155, 157
no ideal rule condition, pattern-type   21,
    22–6, 29–30, 38 n.9
normative–descriptive relationship
    212–13
normative theories   212–13, 231–2,
    238 n.13
    of causal reasoning   206–10
    of verbal probes   235

normativity, sources of   207, 210–12
nudges   152, 162–5, 163–4, 166, 166 n.4
numerically driven inference paradigm (NDI)   72

observation   133
omissions, environmental perceptions   99
Oppenheim, P.   9
ordinary scientific inquiry, and novel sui generis epistemological conclusions   136–7
"our" concept of causation   207–8, 209, 216, 232, 233, 234, 240 n.22

Paris climate change accord, US withdrawal from   66
patterns   15–16
  ideal rule and no ideal rule conditions   21, 22–6, 29–30, 38 n.9
  simple and exceptionless   16, 17–20
Paul, L.   214
pedagogical cues   132–3
personal control   80, 81
Pesendorfer, W.   157
Pew Research Center   97
philosophers
  adherence to innateness concept   174–5
  judgments of   239 n.18
  of science   1–2
philosophy of economics, observable/unobservable distinction to   150
physicists vs. nonphysicists   102–3
physics, errors of commission   99–100
Plato   172
Põder, K.   164
polarization   71, 75–6, 79, 82, 86, 87, 88
political ethology   62, 89 see also nationalism
positive program   2, 236, 240 n.22
preexisting theories, modifications of   180–1
preference-based choice theories, and agent's behavior explanation   158–9
preference concepts   150–1, 155–61, 165, 167 n.6
  and causal bases   158

and choice theory   158
  conceptual analyses of   157
  conceptual changes   158
  constructive concepts   160–1
  and domain of choice theory application   158
  mentalistic interpretation of   157–8
  mental vs. behavioral interpretations of   157–8
  as subjective total evaluations of alternatives   164–5
price change   152
Prinz, J. J.   174
prompt
  to explain   16–17, 18, 20, 21, 22, 25–9, 30, 37–8 n.9
  to write   19–20, 22–3, 25, 26, 29, 30, 37–8 n.9
pro-nationalism   78, 81

Quinean dictum   135–6

Ranney, Michael   10, 64, 70, 72, 73–4, 75, 82, 83, 87, 91 n.9
rationalism   64–6, 90 n.2
rationality concept, in economics   162
realism   150–1
reasoning   135–6, 137
  and acquisition of accurate beliefs, scientific explanations of   136
  causal reasoning (see causal reasoning)
  by children   126–7, 132
  explicit reasoning   210
  a priori reasoning   238 n.13
reason-superiority   65, 71
reference   2
reflective equilibrium strategy   239 n.17
Rehder, B.   18
Reinforced Theistic Manifest Destiny (RTMD) theory   63, 64, 66, 67, 76–7, 87
  as a causal model   77–82
  as a Millian theory   62–4
Reinholz, D.   83
religious beliefs   81
representative scientific knowledge   72, 74
  and global warming acceptance   87, 88
Richerson, P. J.   174

Ross, D. 151–2, 160, 161, 163, 166, 167 n.9
rule ratings 22, 24, 26–7
rule reporting 23, 28
    better rule 21–2, 23–6, 29–30
    worse rule 21, 22–3, 24, 26, 27, 30
Rutjens, B. T. 81

Samuelson, Paul 156
Savage, L. J. 156
Schaffer, J. 219
Schulz, L. 111
scientific ideas, difficulty in grasping 98
scientific knowledge improvement, implications for 113–15
scientific literacy, of general public 97–8
scientific realism 150
scientists
    adherence to innateness concept 174–5
    concept of innateness 179–80
    effective concept 197
    explicit concepts 196
    understanding, and students' understanding 114
    variation in opinions about the importance of innateness 195
    vs. nonscientists 100, 101, 103, 104–5, 115
selective evidential filtering 134
Shanks, D. 229
shared practices of judgment 216, 218–19
Shtulman, Andrew 111
Sloman, S. 219
speciation 108
species-typical traits 173
Sperber, D. 181
spontaneous generation 107
stasis/science-impervious theory, disconfirmation of 65, 71, 82, 87
Stich, S. 239 n.20
Stotz, Karola 7, 180
students
    beliefs, compared with yesterday's scientists 100
    knowledge, compared with scientists' knowledge 98
    preinstructional beliefs 100
success 231
    in connection with causal cognition 209

successful causal analysis and a survey with covariation, difference between 222
super-nationalism 76–82, 88
supra-nationalism 76–82, 86, 88
surprise ratings 75, 80
surveys 4
    bare surveys 217–19, 240 n.21
    and causal analysis, distinguishing between 223–4
    with covariation 219–20, 222
    -like experimental philosophy 236, 240 n.21
system of practice 162

table of denial 88
teleology 176, 177, 178, 179, 180, 183, 185–6, 188–90, 193–4
Tenenbaum, J. 206
testimony 127
theories
    and abstraction 125–8
    cognitive functions of 125–6
    domain-specific theories 123
    linking judgments with contexts and goals 126–8
theory development in history of science 112
theory of evidence 10, 122, 123, 125–8, 126, 139 n.1
    abstract concepts 127–8
    accuracy monitoring 129–30
    determination of information relevant to learning 139 n.2
    inferential principles 128
    and judgments 128
    knowledge and ignorance, distinguishing between 130–1
    and learning 123
    and theory of mind 126
theory of mind 123, 125
three-factor model of animal natures 173, 176–7, 179–80
    fixity 173, 177–8, 180, 182, 183, 185, 186, 188–90, 193, 194–5, 197–8
    teleology 176, 177, 178, 179, 180, 183, 185–6, 188–90, 193–4
    typicality 172, 173, 177–8, 180, 182–3, 185–6, 188–9, 193, 194–5, 197–8
Tobia, K. 239 n.20

traits 111, 172–3, 197–8
　acquired traits 172
　analogous traits 108, 113
　behavioral trait 111
　biological traits 173, 176
　developmental fixity 173
　external anatomical trait 111
　functional traits 173
　homologous traits 108
　internal anatomical trait 111
　species-typical traits 173
truth values 52
"turn to practice" 6–7
Turri, J. 218
typicality 172, 173, 177–8, 180, 182–3, 185–6, 188–9, 193, 194–5, 197–8

understanding, implications for 113–15
United States, nationalism and global warming acceptance in 77

Urban, J. 88
utility theory 90, 150–1, 155, 156, 157

van der Linden, S. 65
van der Pligt, J. 81
Vasilyeva, N. 221, 241 n.35
Velautham, Leela 92 n.10
vitalism 99
voluntariness 155
Voss, James F. 62

Walsh, C. 219, 222
Wegener's theory 105
well-adaptedness, and causal cognition 209
Wilkenfeld, D. A. 31
Williams, J. J. 16, 18, 37 n.6
Woodward, James 8, 222, 240 n.25
worse rule reporting 21, 22–3, 24, 26, 27, 30

www.ingramcontent.com/pod-product-compliance
Lightning Source LLC
Chambersburg PA
CBHW050324020526

44117CB00031B/1747